Hairy Hippies and Bloody Butchers

Protest, Culture and Society

Protest movements have been recognized as significant contributors to processes of political participation and transformations of culture and value systems, as well as to the development of both a national and transnational civil society.

This series brings together the various innovative approaches to phenomena of social change, protest and dissent which have emerged in recent years, from an interdisciplinary perspective. It contextualizes social protest and cultures of dissent in larger political processes and socio-cultural transformations by examining the influence of historical trajectories and the response of various segments of society, political and legal institutions on a national and international level. In doing so, the series offers a more comprehensive and multi-dimensional view of historical and cultural change in the twentieth and twenty-first centuries.

For a full volume listing, please see back matter

Hairy Hippies and Bloody Butchers

The Greenpeace Anti-Whaling Campaign in Norway

Juliane Riese

berghahn
NEW YORK • OXFORD
www.berghahnbooks.com

First published in 2017 by
Berghahn Books
www.berghahnbooks.com

Library of Congress Cataloging-in-Publication Data

A C.I.P. cataloging record is available from the Library of Congress

British Library Cataloguing in Publication Data

A catalogue record for this book is available from the British Library

ISBN 978-1-78533-528-0 hardback
ISBN 978-1-78533-529-7 ebook

Contents

Figures

Acknowledgments

I am very grateful to all current and former Greenpeacers who acted as my informants when I was researching the Greenpeace campaign against Norwegian whaling. Thank you for your trust and your openness. Without you, this book obviously could not have been written.

Many thanks to the editors of the Protest, Culture and Society series Martin Klimke, Joachim Scharloth and Kathrin Fahlenbrach and to the two reviewers who read earlier versions of the manuscript, for extensive critical and constructive feedback and support. I am equally grateful to the wonderful staff at Berghahn Books, particularly Chris Chappell, for their help and patience. I could not have wished for a better reviewing and publishing process than this one. Any errors in this book are my own.

I would like to thank Professors Georg Müller-Christ, Tore Bakken, Dorle Drackle and Adelheid Biesecker for their great support while I pursued my doctoral research. Special thanks to Lars Arndt, who read and commented on earlier versions of this work, and helped me many times to get on, and stay on, the right track. Thanks to Oliver Dehne for research assistance.

I would like to thank the journal *Soziale Systeme* and the *Journal of Applied Behavioral Science*, as well as Berghahn Books, for giving me permission to include parts of earlier publications in this book. I would further like to thank the staff at Østfold University College's library in Halden, who are fantastically helpful.

Many thanks to Dick de Gilder, Naomi Ellemers, Katharina Jantzen, Dorothee Konings, Jutta and Bernd Meyer, Anouk van Leeuwen, Erik vom Hövel and Vigdis Lunde Wisur for their friendship and their help. Finally, I would like to thank Christoph Manuel Meyer for turning my handwritten sketches into printable digital figures, and for his love and support.

Introduction

Observing Greenpeace through the Systems-theoretic Lens

> I think a lot of stuff that we do is not the things that we're campaigning about. It's about changing the world, the way people feel about their place in society, and what they can do, what they can achieve.
>
> When I was young I wanted to work for Greenpeace because they changed things. Gave people hope. Gave me hope. To think that I wasn't a nonexistent person who had to wait every five years to vote for candidates I didn't care about. Showed that you could have an influence as an individual or a group of individuals. I thought there was an organization that meant something, and allowed me to become more meaningful as a part of society, allowed me to say things I wouldn't be able to say otherwise.
>
> —John, Greenpeace Fundraising Director

Few protest campaigns by individual social movement organizations have been as deeply engrained in the global collective consciousness as the Greenpeace campaign against whaling, with its pictures of tiny inflatables going between the ready-to-fire harpoons and the fleeing whales. Greenpeace started to undertake direct actions against whaling ships in 1975. The protests of Greenpeace and many other organizations helped to raise enormous awareness of whaling around the globe and to mobilize international pressure on whaling nations to stop the practice. The argument can be made that the international protests were successful on many fronts. In 1986, a moratorium on commercial whaling decided by the International Whaling Commission (IWC) became effective. It is still in force at the time of writing. Iceland gave up commercial whaling from 1986 until 2006. In

Norway, too, there was no commercial whaling during the period 1988–1992. Japan only conducts whaling for purposes of 'scientific research'.

However, depending on the choice of perspective, it is also possible to make the reverse argument that the anti-whaling protests were unsuccessful in many respects (see Giugni 1998 on the difficulty of assessing the outcomes of protest). At the time of writing, Japan, Iceland and Norway were still allowing whaling or had allowed it to resume. Communication between the global anti-whaling and pro-whaling camps in general and between these two camps in the IWC in particular appears to be deadlocked (Kalland 2012; see also Dorsey 2013).

The Greenpeace campaign against whaling in Norway may be considered a prime example of unsuccessful anti-whaling protest. As I will explain, Greenpeace (and others campaigning against whaling) failed to convince Norwegians that whaling should be stopped, as the anti-whaling campaign was anathema to Norwegians on grounds of their culture and history. Indeed, in a reversal of Greenpeace's self-perception as environmental Davids fighting the Norwegian whale-butchering Goliaths, Greenpeace and the anti-whaling protest community were perceived by Norwegians as foreign oppressive Goliaths and sentimental hippies attempting to impose cultural imperialism on Norway, the environmental David. The anti-whaling protests provoked a committed and sustained counter-campaign by pro-whaling activists (see for example Kalland 2012), as well as general public resistance in Norway. The conflict between Greenpeace and other anti-whaling activists and their Norwegian opponents became known in Norway as the 'whale war' (Furuly 1993b; Johannessen and Bertinussen 1995; Jonassen 1993; Mathismoen 1992a).

In the late 1990s, Greenpeace attempted to redress this deadlocked situation through an in-depth analysis of and a concerted strategy change in its whale campaign in Norway. This book describes Greenpeace's public activities in the whale campaign in Norway from the campaign's beginnings in the late 1970s and the 1980s until 2006, and analyses the organization-internal dynamics behind the campaign. The book's particular focus is on the processes of organizational 'stuckness' of the early to mid-1990s and the processes of organizational reflection and strategy change of the late 1990s.

While the histories of Greenpeace and of whaling in Norway are interesting in and of themselves (see Dorsey 2013; Zelko 2013), the kind of research presented in this book is also important because it can increase our understanding of why and how social movement organizations succeed in achieving social change, or why and how they fail to do so. It can increase our understanding of the conditions and causes of the outcomes of protest. This is a subject we know too little about (Giugni 1998; Louis 2009). We do know that while the social context of protest influences what protest can achieve, the specific activities of protesters also contribute to the success or failure of protest. What protesters do matters for what they can accomplish (Gamson 1975; Giugni 1999). In particular, it matters how they organize, make collective decisions, keep their collective action going. Hence, theorists and practitioners alike can learn about protest effectiveness from studying the internal dynamics of protest (McAdam et al. 1988; Banaszak 1996; see also Tilly 1999). And they can learn at least as much from analysing the organizing mistakes as from the success stories (see also Giugni 1999). Yet the mistakes of social movement organizations such as Greenpeace have rarely been the focus of movement analysts (Minkoff and McCarthy 2005).

Protesters' organizing and decisions – and their mistakes – are influenced by their values, their identities and their perceptions (Banaszak 1996; Van Zomeren et al. 2008). Thus, we need to understand these values, identities and perceptions in order to understand the success and failure of protest. In particular, we need to understand the 'blind spots' of protesters and protest organizations, their 'inability to hear or understand what others are saying' (Mansbridge 1986: 118, 191) that come with their values and identities. In-depth case studies of individual social movement organizations and campaigns are a good research strategy for this, as they can provide us with detailed, contextualized, 'thick' descriptions (Geertz 1973)[1] and interpretations of the organizational processes behind protest. They can inform us about the meaning protesters produce (see for example Yin 1994; Stake 1995; Tilly 2006).[2]

This book analyses Greenpeace's organizational dynamics in the history of its campaign against whaling in Norway. It seeks to understand the values, identities and perceptions, blind spots and inability to hear or understand what others were saying, that influenced these organizational dynamics. It also analyses Greenpeace's organization-internal reflection about all of these and Greenpeace's attempts to improve the situation in Norway. I hope that it will contribute towards an increased understanding of social movement organizations' effectiveness (or lack thereof) in achieving social change.

I witnessed part of Greenpeace's processes of organizational reflection and change myself, as I worked as a full-time volunteer for Greenpeace in Scandinavia from the summer of 1998 to the summer of 1999. Years later, during my Ph.D. trajectory and the process of writing this book, I conducted twenty-five semi-structured in-depth interviews about the Greenpeace whale campaign with twenty-two current and former employees of Greenpeace in Scandinavia, from activists to Executive Directors.[3] I also analysed internal and public Greenpeace materials, articles from newspapers in Norway and other countries, and different websites. I was thus able to develop a case narrative and interpretation. (For a more detailed account of the research process and methodology, see Riese 2015.)

In this book, I hope to have faithfully represented, and done justice to, the differential perspectives of my informants. My interviewees are not quoted under their real names. I do not quote all individuals I interviewed, as not all of them gave me permission to do so. To improve readability, direct quotations from the interview transcripts have been shortened and edited for verbal tics, grammatical and syntactical errors, etc., that are acceptable in conversation but would have made them difficult to read.

I analyse Greenpeace's situation in Norway and Greenpeace's organization-internal processes from a systems-theoretic perspective. In particular, I use the systems-theoretic conceptual framework of 'double bind' to describe how Greenpeace manoeuvred itself into a situation in Norway where, whatever it did, it was always bound to lose – and how it later tried to get out of that situation. In addition to the double bind framework, I draw mainly on the theory of autopoietic social systems as developed by Niklas Luhmann (1927–1998), a German sociologist.

Application of this type of systems theory in the field of protest and social movement research has been limited (but see Ahlemeyer 1995; Luhmann 1997c). This is regrettable, since this theory has great explanatory power and solves various theoretical problems other theories have struggled with (see for example Bakken and Hernes 2003b). In particular, the theory is useful for explaining organizational processes and dynamics (see for example Luhmann 2006; Bakken and Hernes 2003a; Seidl and Becker 2005b). The contribution it can make to the study of protest is potentially great (Hellmann 1998, 2000; Ahlemeyer 1995).

The problem with the plan of using Luhmann's theory of autopoietic social systems to analyse the Greenpeace whale campaign in Norway is that, as Seidl and Becker (2005a: 10) put it, when starting to read Luhmann it takes one or two hundred pages before one understands anything. It is a highly abstract theory which draws on Luhmann's encyclopedic knowledge of the theorizing of thinkers such as Husserl, Parsons, Spencer-Brown, etc. – we might call it a theorists' theory. To begin to understand how the theory might help one to understand social realities, one must then proceed to read hundreds of pages more. (In German.) Consequently, there is a good chance that a reader who has never studied Luhmann before will find a comparatively short and condensed exposition of his theory gruelling and hard to understand. It should be noted that empirical application of Luhmann's theory has generally been limited so far, which is probably an indicator of how challenging an undertaking it is perceived to be (Becker and Seidl 2007;

Vos 2005; but see Nassehi and Nollmann 2004, who say that Luhmann's theory is actually very empirical).

Furthermore, it has to be said that Luhmann is not exactly uncontroversial, whether in his own field or in other disciplines. His theory provokes quite fierce reactions, both of approval and rejection, which is surely due in part to its abstractness and complexity, but also to its highly idiosyncratic character. One reason why its application in protest and social movement research has been limited may be that it is – in my view, wrongly – suspected to be in favour of (structural) conservatism and the status quo (see Hellmann 2000; Rucht and Roth 1992). Another may be that Luhmann comes across, in Nassehi and Nollmann's (2004) words, as an 'aloof ironist' (p. 16, my translation). This possibly makes him less appealing to social movement researchers than for example Bourdieu, who much more than Luhmann is seen, and saw himself, as a political intellectual and political fighter. (As an example of Luhmann being unappealing to social movement researchers, see Luhmann 1988.) The double bind framework is not uncontroversial either (see for example Olson 1972; Putnam 1986).

I do not wish the readership of this book to be limited to systems theory aficionados. In order to make the Greenpeace Norway case accessible to a diverse audience, I have chosen a limited number of systems-theoretic ideas to include and use in the main part of this book, and try to present these in as intuitive and accessible a fashion as possible. The interested reader will find some additional systems-theoretic explanations in the appendix. Choosing this strategy means that I cannot convey the complexity of the theory (let alone discuss its historical development, critical reception, etc.). I hope, however, that the ideas I do include, in the form I include them, will help readers, like they helped me, to understand the case of Greenpeace in Norway.

So which ideas are these? Niklas Luhmann was strongly influenced by the work of the biologists Maturana and Varela, who researched the functioning of the human nervous system (Hernes and Bakken 2003: 13f.). One of their main points is based on the following fact: it can be demonstrated that there is no unambiguous correlation between the wavelength of the colour of an object that we see and the activity pattern or state of the neural system. Meanwhile, a particular state of the neural system does correspond with the name we give a colour. Maturana's and Varela's message is that how we see an object is not determined by the characteristics of that object in itself. How we see an object – our experience – is determined by our own structure of cognition. The neural system is structurally determined (Maturana and Varela 1987: 26f., Ch. 7; Bakken 2000: 64ff.; Simon 2000a: Ch. 4).

This means that we should not think of cognition as an input-output model (as in, the environment gives an open system information). We should use a concept of operationally closed systems (Maturana and Varela 1987: Ch. 6, 176ff.; Bakken 2000: 74f.; Simon 2000a: 79). Maturana and Varela used the term autopoiesis (Greek: autos = self, poiein = to make) to describe living systems characterized by the ability to produce and reproduce the elements they consist of. An autopoietic system is operationally closed because the operations leading to the productions of new elements in the system are dependent on earlier operations of the system and are the basis for the following operations (Maturana and Varela 1987: 50ff.; Maturana 1999: 153f.; Baraldi, Corsi and Esposito 1997: 29f.; Luhmann 1997b: 65ff.). States of the nervous system trigger other states of the nervous system. These states cannot be produced outside of the system and then 'transferred' into the system. Neither can the system transfer its states into another system. If it is no longer possible for the nervous system to connect one state to another, the system will cease to exist (Maturana and Varela 1987: 186f.; Simon 2000a: 78ff.).

Luhmann generalizes the concept of autopoiesis and applies it to psychic and, most importantly, social systems (Luhmann 1997b: 66; Baraldi, Corsi and Esposito 1997: 29f.). In the Luhmannian framework, both psychic systems and social systems are understood to produce their own states, or elements, and connect one element to the next one, and then to the next one, and so forth. The elements of a psychic or social system cannot be produced outside of that particular system and then transferred into the system. Neither can a psychic or social system transfer its elements into another system. The elements of psychic systems are experiences; the elements of social systems, communications.

Information in social and psychic systems, then, is a system-internal quality. In the case of an autopoietic system, information cannot be in the system's environment and then be transferred into the system as an 'input' to which the system is open. For example, a psychic system cannot 'give' information to another psychic system. Neither can it 'give' information to a social system. Instead, information depends on the system in which it is processed. It is produced in the system. In the tradition of Bateson, systems theory scholars say that information is 'a difference which makes a difference' (Luhmann 1995b: 40; Bateson 1983b: 582). If something makes a difference in a system, it is information for that system. On the other hand, if something does not make a difference in a system, then it is not information for that system (Luhmann 1990: 45). For example, when you hear people talk in a language you do not understand, their talk is simply white noise for you and not information. It does not make a difference to you what noises they make, because you don't understand any of the noises anyway.

While psychic and social systems are operationally closed, they are cognitively open. External stimuli can serve as perturbations to a psychic or social system, which are then processed by the cognitive structure of the system. The metaphor of a kaleidoscope is apt here (Simon 2000a): an external stimulus (a shake, an impact) can induce the kaleidoscope to rearrange its internal colourful structures. But this rearranging takes place under the terms of the kaleidoscope, it is done by the kaleidoscope itself. If one tried to directly arrange the coloured bits and pieces inside the kaleidoscope, one would in effect destroy it. An autopoietic system can and does realize and process perturbations from outside, but on its own terms.

This means that psychic and social systems can and do influence each other, but in the sense of mutual perturbation, not in a sense of mutual determination. If we take Luhmann's theory as a working basis, we become sceptical of our everyday notion that individuals may 'steer' or even 'force' social systems to operate in a certain way (as in, the boss is at the wheel of the company). Social systems possess autonomy; they possess *eigen*-dynamics which cannot be determined by psychic systems. (Neither can social systems control psychic systems.) A member of an organization may have ideas on how to improve the work of the organization, and she may try to make the organization adopt those ideas. But if the organization doesn't buy them, doesn't get them, doesn't adopt them on its own terms, then her efforts will be futile. Greenpeace in Norway illustrates this neatly: Norwegian Greenpeacers did see, and did try to make their organization see, that Greenpeace's campaigning in Norway was counterproductive. But for a long time, they could not get their organization to understand this.

In a unique way, Luhmann's theory acknowledges research findings which make it clear that it is impossible to explain social events by referring to mental states of actors. The social situation is decisive for how individuals 'act' or, indeed, for what constitutes an 'action' or an 'actor,' or for what an 'action' means (see for example Ellemers 2012 or Ellemers and de Gilder 2012; Luhmann makes this point in Luhmann 1995b: 165f.). Interestingly, Luhmann's theory was eagerly taken up by family therapists (Simon 2000b). Family therapists know that the thoughts and needs of individuals often, mysteriously, fail to be communicated in the social system that is the family. They also know that dysfunctional patterns in families often seem to emerge against the will or 'behind the backs' of the family members. The longer one looks at a particular family, the more impossible it becomes to pinpoint who started a particular pattern, or who is causing a particular problem. Thus, in a unique way, Luhmann's theory frees us from having to attribute responsibility for social events to individuals and their supposed motives (although Luhmann is clear that such processes of attribution are very much a part of our lives). It frees us from having to play the blame game.

My interpretation of Greenpeace's problems in Norway revolves to a considerable degree around the problem of internal 'representation' of the environment in operationally closed systems. Because social systems cannot receive direct informational input from their environment, they can never know what their environment 'is really like' (Luhmann 1995b: 34f.). (This idea is by no means exclusive to Luhmann. See for example Burr 2003: Ch. 5.) Instead, in the process of (system-internal) cognition, the social system, on the basis of the external perturbations it processes, produces an internal under-complex reconstruction of what is outside of it. The reconstruction must be under-complex, because the system is less complex than the environment. If the reconstruction were complete – if there were a point-to-point correspondence between system and environment – the system would have to be as complex as its environment, and then it would be pointless to speak of a system. (Luhmann does not even accept the term representation instead of reconstruction, for even this would be too hopeful; Luhmann 1984: 47ff.; Luhmann 2006: 314; Luhmann 1997b: 124.)

It is intuitively understandable that it is decisive just which differences make a difference for the system, what exactly is information for the system (and what is merely white noise). What if the system misses crucial points about its environment? What if its internal reconstruction of its environment is under-complex in a bad way, such that it does not include things that are important for the continued reproduction – in other words, existence – of the system (Luhmann 1984: 47ff.)? What makes such situations particularly dangerous and difficult is the fact that a social system can miss a point *and* fail to understand that it has missed a point. In other words, a social system can miss environmental perturbations that indicate it is not picking up important environmental perturbations. (The analogous argument applies to psychic systems.) As we shall see, this is exactly what happened to Greenpeace in Norway.

The fact that this resulted in a notable lack of success for Greenpeace in Norway for a long time did not really make the organization reconsider its strategy (at least not in a comprehensive fashion). It is often taken for granted that if a social system does not reach its goals, the social system will reflect on why not, and change its strategy. Not so, says Luhmann: purpose is not a sufficient guide for action or reflection. Building on March and Simon (1958: 165), Luhmann says that purposes, such as stopping whaling, are not motives which allow one to understand and explain organizations' operations, although organizations may use a purpose as a justification for certain decision chains.[4] The main 'purpose' or 'goal', the first priority, of an autopoietic system such as an organization is to continue operating, maintaining its own autopoiesis (Luhmann 2006: 165, 183–185, 256f.). What Greenpeace did in Norway did serve the function of enabling Greenpeace's own autopoiesis. It meant that Greenpeace could continue operating, and it mobilized support for Greenpeace in many countries (although not in Norway). So the fact that it did not reach its goal did not suffice to make the social system Greenpeace change.

It should be noted again at this point that Luhmann does not explain social autopoiesis in terms of motives or thoughts of individual actors (psychic systems). The fact that the main 'purpose' of an organization is to reproduce itself is not explained for example by the wish of its members not to lose their jobs in the organization. Instead it is seen as being similar to a human body (also an autopoietic system) continuously renewing its cells and so on, i.e. continuing its own autopoiesis, without any particular 'motive' or 'goal' except continued existence. It will be seen that this applies to Greenpeace in Norway, as well. That Greenpeace did not win the whale campaign mattered enormously to all Greenpeacers, both Norwegian and international. They would have liked to win the whale campaign all along. Also, most of them would have been able to find a good job with a different organization (and many eventually did). Greenpeacers' individual motives and interests are not a sufficient explanation for the social system Greenpeace's inability to change its whale campaign strategy.

According to Luhmann, reflection happens when, instead of unquestioningly continuing to reproduce itself, the system observes itself as a contingent unity in an environment. Something is called contingent when, to put it very simply, it could be otherwise. Contingent is 'neither necessary nor impossible' (Luhmann 1984: 152; Luhmann 1995b: 106). When observing itself as a contingent unity in an environment, the system is, potentially, able to compare this unity with alternatives. It is, potentially, able to realize whether it has failed to pick up on important environmental perturbations, and to choose to operate differently in the future so as to take these into account.

It must be emphasized that reflection is a special achievement, not something we can expect to happen all the time (Luhmann 1984: 601f., 617ff.; Baraldi, Corsi and Esposito 1997: 154f.). For an autopoietic system, whose main 'goal' is to maintain its own autopoiesis, the next step is typically more important than the future, because without the next step it will not reach the future (Luhmann 1990: 38; Luhmann 2006: 53). So it mostly just goes on reproducing without reflecting on its reproduction. Strictly speaking, social systems don't need reflection – as long as their operations work well enough for them to survive (Vos 2005: 375). And when their operations no longer work well enough for them to survive, it is often too late for reflection anyway.

If and when a system reflects, uncertainty is increased. Reflection makes life harder for a social system, because it produces awareness that the system's structure is contingent, that the system could communicate differently, that it might have failed to process important environmental perturbations. The analogous argument applies on the level of the psychic system: an individual who consciously thinks about every step she takes and wonders whether she should have taken a different step will find it hard to walk. In this sense, it is healthy that social systems (and psychic systems) do not reflect all the time.

In reflection processes of social systems, such as Greenpeace's, emotions play an important role.[5] This is because emotions can be signals indicating the viability of a social system, its 'fit' with the environment.[6] Emotions may signal that the structure of the social system is viable in the environment, in other words, that the social system is doing well and should continue like this. In other cases, emotions may signal that the social system is threatened in some way (cf. Ciompi 2004). They may then trigger contradiction in the social system.[7] Luhmann speaks of the immune function of contradiction. Contradiction can help the social system to protect itself with the help of changes 'against rigidifying into repeated, but no longer environmentally adequate, patterns of behavior' (Luhmann 1995b: 371f.).[8]

A social system which wants to reflect on its own structure – an organization undergoing planned change, or a family in a therapeutic setting – must take into account the emotions the system's members experience. Emotions signify 'reasons' why the current structure is adequate in the environment. But they also signify potential dangers, 'reasons' why the structure is no longer sustainable. A social system which is able to communicate about its members' emotions, and to create an awareness of them, may succeed at reflecting on potential gaps between its actual structure and the structure that would be necessary. In consequence, it may be able to change purposefully for the better. On the other hand, a social system can waste a lot of time and resources talking about the 'factual' side of problems if it doesn't pay attention to the 'underlying' emotions (Günther 2004). Family therapists and organizational consultants have of course long known this (see for example Kahn 2003; Stein 2001; Vince and Broussine 1996; Watzlawick 1978). This book thus proceeds from the assumption that in order to behave 'rationally,' an organization must be 'emotional' (see also Carr 2001).[9] Further systems-theoretic concepts will be explained in subsequent chapters.

The book is structured as follows. In Chapter 1, I account for the early history of Greenpeace and provide a short introduction to Norwegian history and culture. The dynamics of the 'whale war' between Greenpeace and Norwegians can only be understood against the background of these histories. When Greenpeace protested against whaling in Norway, Greenpeace's organizational self-description clashed with the dominant Norwegian self-description. Because both Greenpeacers and Norwegians were strongly emotionally attached to their respective self-descriptions, the confrontations between the two groups became highly emotionally charged.

Chapter 2 offers a short overview of the history of Greenpeace Sweden and Greenpeace Denmark before the two organizations merged with Greenpeace Finland and Greenpeace Norway to form Greenpeace Nordic. Greenpeace in Sweden, before the merger into Greenpeace Nordic, was the organization which first adopted the approach that came to be known as 'ultimate campaigning' or '*Phyllis Cormack* campaigning'. This approach was then brought to bear on Greenpeace's anti-whaling campaign in Norway in 1998–1999. Greenpeace in Denmark engaged in serious conflict with the umbrella organization Greenpeace International (based in Amsterdam) over a campaign in 1996. This experience became relevant when Greenpeace Nordic had to convince Greenpeace International of its new strategy for the Norwegian anti-whaling campaign in 1998–1999.

In Chapter 3, I describe the history of Greenpeace's anti-whaling campaign in Norway from 1988, the year Greenpeace opened its Norwegian office, until the merger into Greenpeace Nordic. This history was characterized by what I call 'loops of chaotic communication': the 'David against Goliath' chaos communication loop, the external 'campaigning against each other' chaos communication loop, and the internal 'campaigning against each other' chaos communication loop. These loops were unproductive, but unfortunately also self-reproducing and self-reinforcing, patterns of communication that Greenpeace was unable to get out of. I explain that Greenpeace Norway got caught in a pathogenic organizational double bind, a situation where it was 'damned whatever it did'. This organizational dynamics was due, to a considerable extent, to the fact that the organization was emotionally attached to its self-descriptions.

Chapter 4 first describes the process of merging the Greenpeace offices in Finland, Norway, Sweden and Denmark that took place between 1997 and 1999, and the resulting regional organization. I then account for Greenpeace's organizational reflection about the campaign against Norwegian whaling and about the related organizational self-descriptions, a reflection that led to a strategy change in the campaign in 1999. In particular, I describe how two Greenpeacers I will call Nils and John acted as change agents and put their organization in a therapeutic organizational double bind. Therapeutic organizational double bind is a method of intervention which is employed in order to enable an organization to resolve the pathogenic organizational double bind it is caught in.

Chapter 5 explains how, after Nils and John left Greenpeace Nordic in 2000, the organization relapsed into the pathogenic organizational double bind that characterized the campaign against Norwegian whaling. I conclude with some reflections about reflection in protest organizations. The appendix offers some additional systems-theoretic explanations.

Notes

1. Geertz indicates that he borrowed the notion of 'thick description' from Gilbert Ryle.
2. Ideally, we should compare systematically across cases, movements and time (Benford 1997; Minkoff and McCarthy 2005); but of course an in-depth understanding of individual cases is a prerequisite for this.
3. The interviews took place in 2005 and 2006. Interviews with Swedes were done in Swedish, and translated into English by me. All other interviews were conducted in English.
4. Luhmann (2006: 165) quotes March's idea that organizations are systems that search for purposes (see March and Olsen 1976).
5. The theoretical ideas on emotions included here were earlier published in Riese, J. 2011. 'Functions, Communication, and Perception of Emotions in Luhmannian Theory: Emotions as Reflection Resources of Social Systems', *Soziale Systeme. Zeitschrift für soziologische Theorie* 17(1): 53–72. Republished with permission.
6. Luhmann regards emotions strictly as a psychic phenomenon (Luhmann 1984: 370ff.; cf. Baecker 2004: 10). A social system cannot have emotions (cf. Simon 2004). However, because humans are social animals who find it difficult to live and survive in social isolation, in other words, who depend on 'their' social systems for their own autopoiesis, human psychic systems will develop emotional reactions concerning the perceived degree of viability of the social system.
7. This argumentation is compatible with Ciompi's (2004) *Affektlogik*: Ciompi says that emotional energies can organize a social system in a certain way, until increasing emotional tensions provoke an abrupt bifurcation, a switch to a different pattern when it is no longer possible to cope with a situation in 'the usual way'.
8. Emotions can become relevant or processable in the social sphere in two ways. Firstly, communication can communicate about emotions; emotions can become information in communication. This holds true both for verbal and nonverbal communication. Secondly, emotions can 'show' themselves without being communicated about; we can 'sense' them in others (cf. Simon 2004). Emotions can become information in another's psychic system as a result of perception without any communication having happened. Following Weinbach (2004a, 2004b), I suggest using the – originally Bourdieusian – term 'habitus' to denote this pathway for emotions to become relevant in the social sphere (see Bourdieu 1990, 2001). It is reasonable to think that the perception of emotion via habitus is 'fuller', closer to

the 'real' emotion, than 'mere' communication about an emotion. Seeing an emotion in others might induce us to empathize, mirror the feeling, relate to it. Perception of emotions also has an advantage over communication of emotions with respect to speed and immediacy.

9. Goodwin, Jasper and Polletta note in their aptly titled contribution 'The Return of the Repressed' (2000) that much of the social movement literature is characterized by a 'cognitive bent' (see also Benford 1997). Emotions are often equated with irrationality; it is assumed that emotions and rationality are incompatible (Aminzade and McAdam 2002). The social movement literature shares its 'cognitive bent', which it is beginning to redress, with Luhmann's work. This book's systems-theoretic framework seeks to adequately appreciate the role of emotions. The fact that it uses the habitus concept to do so relates it to contributions such as Haluza-DeLay (2008) and Medvetz (2006), which employ the concept to analyse social movements. Crossley (2003) in particular notes the usefulness of Bourdieu's theory of practice for analysing social movements, and for analysing reflexivity in social movements.

Chapter 1

Antecedents

Greenpeace, Norway and Whales before the Greenpeace Whale Campaign in Norway

The reasons for the 'whale wars' between Greenpeace and Norwegians and for Greenpeace's lack of success in Norway can only be understood if one knows at least a little about Greenpeace's and Norwegians' history. This chapter therefore accounts for the antecedents of the Greenpeace whale campaign in Norway.[1]

Greenpeace: Origins

At its inception, Greenpeace was not a professional, hierarchical, environmental nonprofit organization, but a protest campaign rooted in the international peace and environmental movements. Starting in 1965, the U.S. military conducted a series of underground nuclear tests on the island of Amchitka, which belongs to the Aleutians, a U.S.-American island chain situated between Alaska and northeast Russia. The second of these tests, in 1969, provoked outrage and protest in Canada and particularly in Vancouver, the closest major Canadian city to the test site. Many feared that the bomb might trigger an earthquake which in turn might cause a tsunami that would hit Canada's west coast. At this point, some Vancouver-based protesters formed an independent group to mount a campaign against the next Amchitka test, which was scheduled for October 1971. This new protest group was called the Don't Make a Wave Committee (DMWC) (Zelko 2013: 65ff.; Bohlen 2001: Ch. 1; Hunter 2011).

The DMWC decided to sail a boat to the edge of Amchitka's twelve-mile limit, the area under U.S. jurisdiction, daring the U.S. to either abandon the test plans or tow the boat out of the test area, an action that would effectively constitute an act of piracy. The DMWC would work to draw as much international media attention as possible to the boat's voyage. Famously, the protesters were struggling to come up with a name for the boat, until one of them called Irving Stowe, when leaving a meeting in 1970, flashed his usual V sign and said 'Peace,' to which another called Bill Darnell replied, 'Make it a green peace!' (Zelko 2013: 68f.; Bohlen 2001: Ch. 1–2; Hunter 2011).

The *Greenpeace* never made it to Amchitka in 1971, but its voyage did attract considerable media attention and certainly contributed to raising public awareness of the issues it was highlighting (Bohlen 2001: Ch. 2–3; Zelko 2013: 106ff.; Hunter 2011). It also led to the establishment of the new Greenpeace Foundation in 1972 (Zelko 2013: 106ff.). The *Greenpeace*'s crew members represented the ideas and traditions that would influence the Greenpeace organization: 'Quakerism, the radical pacifism of the American peace movement, the New Left and the counterculture, Marshall McLuhan's theories of mass communication, and various strands of environmentalism, all united by a commitment to nonviolent direct action and a shared belief in the revolutionary potential of holistic ecology' (ibid.: 77).

According to Zelko (2013), the U.S. peace movement owes much to Quakers. Quakers were pacifists and committed to nonviolent protest, because they believed that every person had direct access to God and so was a potential channel of truth. Violence against people, therefore, would only suppress love, truth and freedom (p. 13; see also Bigelow 1959: Ch. 3). A Quaker form of protest is 'bearing witness', expressing one's disapproval of an activity and putting moral pressure on the perpetrators simply through one's presence at the scene (Zelko 2013: 13). A group of Quakers started the Committee for Nonviolent Action Against Nuclear Weapons (CNVA) which in 1958 organized a yacht called the *Golden Rule* to sail into a U.S. nuclear test zone. The yacht's crew was stopped in Hawaii and the action never managed to attract much media attention, but it became legendary among Quakers and pacifists nationwide and inspired similar actions in the years to come, one of which, of course, was the *Greenpeace* action (Zelko 2013: 16ff.; Bigelow 1959; Bohlen 2001: 28; Hunter 2011: 25; on the Quaker influence on Greenpeace, see also Carmin and Balser 2002).

Those who were active in the peace movement were influenced by the horrors of the Second World War and the Holocaust and by the fear of nuclear war. Beginning in the 1950s, these fears became entwined with concerns about environmental contamination, due to events such as the discovery of traces of radioactive substance in mother's milk throughout North America (Zelko 2013: 29ff.; Bohlen 2001: Ch. 1). During the 1960s, an increasing number of people both in the U.S. and Canada came to adopt a holistic, ecological, biocentric rather than anthropocentric worldview, and became active in groups such as the Sierra Club (Zelko 2013: 33ff.; Bohlen

2001: Ch. 1; see also Zelko 2007). (Initially, it was thought that the protest against the 1971 nuclear test on Amchitka could be organized by the Sierra Club, until this idea turned out to be impractical; Zelko 2013: 67f.; Bohlen 2001: 27–28.) A holistic ecological perspective also came to be an important part of the counterculture of the 1960s and 1970s, which some individuals who would be influential in the Greenpeace organization represented (Zelko 2013: 44ff.; see also Hunter 1971).

Different individuals who became involved in Greenpeace subscribed to various perspectives and cultural traditions, ranging from Zen Buddhism and the New Left to Gestalt therapy (Zelko 2013: Ch. 1–3; Bohlen 2001: Ch. 1). All of these perspectives called into question the values of mainstream Western society, its 'notions of progress, [economic] growth, and security' guaranteed by the military-industrial complex (Zelko 2013: 52; see also Carmin and Balser 2002). They thus had important aspects in common. However, internal dissension and conflict have been characteristic of Greenpeace from the very beginning (Dale 1996; see also Zelko 2013; Hunter 2011).

One disagreement among the 1971 *Greenpeace* crew was about how to campaign for the necessary change in society most effectively. One faction proceeded from the belief that ecology was the basis of a new form of consciousness, one that implied a 'transformation of the entire working relationship with the natural world' (Hunter 1971: 123). It was necessary to adopt this new form of consciousness in order to avoid environmental catastrophe (ibid.: 121). This notion became connected with the ideas of communication and media theorist Marshall McLuhan (Dale 1996; Zelko 2013), whom Bob Hunter, a *Greenpeace* crew member who would be influential in Greenpeace during the 1970s, called 'our greatest prophet' (Hunter 1971: 221). McLuhan argued that thanks to modern technology and communication media, humans now lived in 'a global village' (McLuhan and Fiore 2001: 63) where distance in terms of time and space between them had lost relevance.[2] People everywhere knew about each other, were involved with and responsible for each other (p. 24). Politics was therefore changing; the mass media audience could be used as a creative, participating force (p. 22).

Greenpeacers like Hunter thought that the new, holistic, ecological consciousness that was expressed by and through the environmental movement (Hunter 1971: 119) could use the system of mass communication, TV, radio and so on, as a 'delivery system' for 'mindbombs'. The objective of firing such 'mindbomb' messages or impressions was 'to change the consciousness of the "enemy", meaning those still in the grip of the old suicidal modes of consciousness' (Hunter 1971: 216f.). Hunter believed that this revolutionary strategy (p. 221) was incredibly powerful: 'Not even a hydrogen bomb can affect so many people at once [as television]' (p. 218). Would-be ecorevolutionaries had to acquire media savvy: 'Most environmental groups are so busy beating their chests in righteous indignation that they don't take the time to find out what makes the media tick. [... But] You've got to prove your case [in the media]' (Hunter in 1977 in an article in *The Province*, quoted in Weyler 2004: 452).

Hunter wanted to create 'a mythology for the environmental movement ... that would resonate with millions of people and help bring about [a] change in mass consciousness' (Zelko 2013: 79; see also Weyler 2004: 228). Other founding members of Greenpeace thought that this was somewhat impractical (Zelko 2013: 51). They agreed that media attention was important for the 1971 anti-nuclear action, but wanted to run 'a sober campaign that would garner the respect of elite groups, such as scientists, politicians, and high-level bureaucrats' (ibid.: 79). About this and other issues, the *Greenpeace* crew fought constantly (Dale 1996: 15; see also Hunter 2011).

Their disputes were but a harbinger of things to come. Dale (1996) quotes a former Greenpeacer as saying that 'Greenpeace has always been ... crisis-ridden and perhaps crisis dependent ... But I think that's a healthy thing, because the organization is trying to internalize various standpoints' (p. 74). Another former Greenpeacer says that Greenpeace is in 'turmoil on an ongoing basis' because it attracts those in the environmental movement who want something done quickly and effectively, and are not the most patient of people (ibid.). Dale concludes that, given the considerable differences of opinion between different Greenpeacers, what is amazing is not that the history of Greenpeace is full of conflict but that the organization has lasted as long as it has (p. 75). As we will see, all of this is as true for Greenpeace in Norway as it is for the rest of Greenpeace history.

Bob Hunter and other representatives of the faction of 'mystics' (Zelko 2013: 211) within Greenpeace went on to make the protection of whales an important part of their campaigning to achieve a 'consciousness revolution'. In 1973, after several years of anti-nuclear campaigning, these people felt that the Greenpeace Foundation, the successor of the Don't Make a Wave Committee, 'was focusing too heavily on the "peace" half of its name and not enough on the "green"' (ibid.: 159; see also Hunter 2011). For them, protecting whales from exploitation by humans constituted a perfect symbol of environmental survival. Instead of using people's fear of nuclear death as a trigger for consciousness revolution, they wanted to appeal to people's reverence for life (Zelko 2013: 161).

The fact that whales and dolphins were the species which these Greenpeacers felt were representative of their values and philosophy can be interpreted as a result of broader cultural developments. These developments involved controversial research on whales and dolphins,[3] films such as *Flipper*, books, aquariums, etc. (Zelko 2013: Ch. 7–8). They led to the creation and widespread acceptance of the notion of what Zelko calls the 'metaphysical whale': 'a sublime, mystical, ecologically harmonious and super-intelligent aquatic being representing a supreme form of power and intelligence rooted in a oneness with nature, a state that humans, in their dangerous and pathetic struggles to conquer the natural world, could never achieve' (Zelko 2012: 104f.; see also Zelko 2013: Ch. 7–8; Kalland 2012: Ch. 1–2). Whales thus symbolized both the 'green' and the 'peace' in Greenpeace. My interviews of 2005 and 2006 showed that Greenpeacers still subscribed to this interpretation thirty years later.

> *Alex*: In a way whales are a symbol for how we treat the planet. If we can't even save the whales, how the fuck are we gonna save the small bugs. You know, it is something that has such a great emotional appeal to people, to everyone, really, except the Norwegians, the Japanese and Icelanders. If the capitalist economics takes over even that, then it's hopeless (*laughs*)…
>
> The other side of it is, if you've ever been in an inflatable next to a whale, it is an incredible feeling. It is like the presence of life. You know, it's just amazing. And so it's also a personalization of this thing that we're all trying to save. And they're also beautiful, supreme and peaceful, … they are a Greenpeace thing.

Greenpeacers developed strong emotions about whales and whaling (see also Hunter 2011). Meanwhile, as Zelko (2013: Ch. 8; 197; 228ff.; 294) makes clear, Greenpeace's whale campaign was burdened with certain philosophical inconsistencies from the beginning. The case for the 1971 anti-nuclear protest voyage to Amchitka had been very clear-cut, the disagreements among those involved with regard to strategy and tactics notwithstanding. In contrast, Greenpeace used arguments against whaling which partly (by no means completely) contradicted each other.

On the one hand, Greenpeace subscribed to a perspective of holistic ecology. From such a perspective, whaling is the wrong thing to do if it critically diminishes whale populations, thereby disturbing the balance of the greater ecosystem. That this has historically been the case with global industrial whaling is not in dispute (Zangl 1999: 160ff.). However, from a holistic ecology perspective, whaling should be considered as acceptable as long as it is sustainable and not disruptive to the ecosystem as a whole. It is here that this perspective has always clashed with Greenpeace's other arguments against whaling, based on the notion of the 'metaphysical whale':

that whales should not be hunted by humans at all, because they are highly intelligent, 'exemplars of ecological virtue and holistic consciousness' (Zelko 2013: 193), and because each whale is an individual worth saving (see Kalland 2012: Ch. 1), etc. Such a notion of whales could only lead to an abolitionist stance, which Greenpeace has indeed taken over the past decades.

According to Zelko, some early Greenpeacers were aware of the inconsistencies in their argumentations. But the point was that they were working to achieve a 'consciousness revolution' to save the natural environment. Animals such as whales and seals could be 'gateway species': because of their appeal to many people, campaigns for their protection might lead to people developing a deeper appreciation for wildlife generally, and so make it easier to save other, less 'extraordinary' species (Zelko 2013: 244, 247). For someone who seeks to attract media attention in order to lob mindbombs at people with the goal of fundamentally changing their perceptions of reality and bringing about a shift in society towards deep ecology, the described inconsistencies might seem relatively unimportant. And using a variety of arguments, ranging from the overexploitation of whale stocks to the intelligence of whales, might enable such a 'consciousness revolutionary' to persuade a greater number of people of his case. Moreover, it is of course easier for someone engaging in detached academic analysis decades after the fact to point out inconsistencies in an argumentation, than it is for the person developing the argumentation in the context of a new, unfolding and quite risky campaign (ibid.: 238f.; 247).

In 1973, the older, sober activists in Greenpeace opposed the idea of the young organization diverting energy from the anti-nuclear cause to the 'soft' issue of whale protection, which might also dilute the power of the by then well-known Greenpeace brand name (ibid.: 174ff.; see also Weyler 2004: 230; Hunter 2011). But from 1974, the anti-whaling campaign became Greenpeace's most prominent one. In Zelko's words, this 'put Greenpeace on a path … to a more broadly conceived form of environmental activism' (Zelko 2013: 160).

The first Greenpeace direct actions against whaling off the North American West coast in 1975, 1976 and 1977 confronted Russian whalers who operated from fleets that included one huge factory ship and several smaller harpoon boats (Weyler 2004: 311ff.; 411ff.; 428ff.; 463ff.). The activists drove their rubber dinghies between the harpoon boats and fleeing whales, putting their own bodies in the line of fire in order to try and stop the whalers from firing their harpoons at the whales, filming and photographing their actions all the while.

The media, particularly the U.S. media, were receptive to the spectacular Greenpeace anti-whaling campaign (Zelko 2013: 225; Hunter 2011). The cultural developments mentioned above had already influenced many people's consciousness to the effect that whales were 'special' animals (Zelko 2013: 222). Also, Greenpeace was confronting industrial whaling here (Weyler 2004: 314; Hunter 2011). This meant that the abovementioned contradictions in Greenpeace's view of whaling were not as noticeable to either Greenpeace or Greenpeace's audiences as they would become at later points in time. Industrial whaling had driven many whale species to the brink of extinction, and was therefore problematic both from the point of view of those who saw whales as 'special' animals and from an ecological perspective. Moreover, because whaling was of no significance to the U.S. economy, and the main whaling nations were the Soviet Union and Japan, 'Cold War overtones' were present in much of the U.S. reporting on the issue (Zelko 2013: 225). This must have made the issue seem even more straightforward to many U.S. citizens (and citizens of U.S. allied nations), with whalers and supporters of whaling as the 'bad guys' and Greenpeace and other anti-whaling protesters as the 'good guys'. As a result of all this, the campaign was very successful as a media event and brought fame to Greenpeace and a boost to the anti-whaling movement (Zelko 2013: 222–225;[4] see also Weyler 2004: 323; Hunter 2011). (Hunter [2011] gives an account of meeting the Dutch captain of a whaling ship in a bar in Australia in 1977. The whaler told Hunter that he loved being at sea, but did not enjoy killing whales. It was his job. The whaler said to Hunter: 'You don't have to remind me … that that's what the Nazis said, too' [p. 436].)

The successful first direct actions against Soviet whaling irreversibly made whales and the whale campaign central to Greenpeace's self-description. Organizational self-descriptions are productions of texts or functional equivalents of texts through which and with the help of which an organization identifies itself (Luhmann 2006). They are self-simplifications. Self-descriptions provide the organization with direction for its operations and a sense of its own unity. They also serve to emphasize an organization's uniqueness (Luhmann 2006: 438; Seidl 2003).[5]

Individuals seek to identify with organizations that possess appealing organizational self-descriptions, because such identification enhances their self-esteem and offers fulfillment and realization of self. This is true for most organizational contexts (Ashforth and Mael 1989; Diamond 1993), but maybe particularly in the context of social movement organizations like Greenpeace (Gamson 1992: 56). Organizational members are emotionally attached to organizational self-descriptions, and this emotional attachment supports the self-descriptions and maintains them over time, so that they can serve their important functions for the organization.

A Greenpeace self-description that was present from the first days of the first anti-nuclear campaign was that of 'David against Goliath' (Zelko 2013). Greenpeacers in a small boat confronting huge vessels could be seen as courageous warriors foiling the destructive plots of the infinitely more powerful. They were successful despite their lack of strength because they used clever strategies and tactics. The main strategy (as explained above) was to expose their opponents' crimes ('bearing witness') and to change people's consciousness via the media. It should be noted that for many Greenpeacers, their organizational self-description and their actions had little to do with heroism. Ben Metcalfe called the early Greenpeace of which he was a part 'an absurd, pathetic little group'; Bob Hunter described the group as 'anti-heroes rather than heroes' (Dale 1996: 15; see also Erwood 2011: 6).

In addition to being about 'Davids against Goliaths', the Greenpeace self-description was one of protectors who, in a peaceful and nonviolent manner, risk their own lives to save the lives of others put in danger by a violent threat (Zelko 2013). The Greenpeace activists who attempted to enter nuclear test zones in the 1970s risked exposure to harmful levels of radiation. Activists David McTaggart and Nigel Ingram were severely beaten up by French military commandos in 1973 when protesting against nuclear tests (McTaggart 1973). Photographer Fernando Pereira was killed when the Greenpeace ship *Rainbow Warrior* was bombed and scuttled by the French Secret Service in 1985 (King 1986).

The first Greenpeace actions against whaling, during which Greenpeacers drove their tiny zodiacs between whaling boats and whales, running the risk of being killed by a harpoon (Weyler 2004: 464f.; Hunter 2011) in order to save whales, were a powerful new manifestation of the Greenpeace self-description. The Quaker- and peace movement-inspired tactic of a group of human beings deliberately putting their lives at risk was used to protect nonhumans. This was revolutionary (Zelko 2013: 222). In one way, it can be interpreted as a manifestation of a biocentric rather than anthropocentric worldview (Zelko 2013: 8). Weyler (2004: 230) quotes Bob Hunter as saying: '[W]e have to shift the human-centred perspective. From a very real point of view, saving whales is as important as saving humans. We shift from the sacredness of human life to the sacredness of all life'. Since these first actions against whaling, the Greenpeace self-description has been inextricably linked to the protection of whales. As one of my Swedish interviewees said in 2005:

> *Nils*: Putting a rubber boat between the harpoon and the whale, it's an incredibly strong communication, if you want to put it like that. A strong picture. So it's obvious that it stuck in the collective consciousness in some way.

> It's several such almost archetypal symbols, partly the small one, David against Goliath, and it's also that you risk your own life to save someone else's, who is weaker, defenceless… which the whale is in that case. … A harpoon, a murder weapon, and then a vulnerable activist … Jesus on the cross… I am not religious myself, but I can imagine that maybe people have these associations.

Until 1979, Greenpeace developed mostly according to an organizational philosophy that Bob Hunter described as 'Let a thousand Greenpeaces bloom' (Zelko 2013: 302; see also Weyler 2004: 404; Hunter 2011). Basically, anyone anywhere in the world was allowed to start a Greenpeace chapter if they wished to do so, and the different groups formed a kind of network rather than a unified organization (on the organizational issues Greenpeace was facing, see also Weyler 2004: 458ff.). Zelko (2013: Ch. 12) describes how this grassroots organization was eventually converted, within a relatively short time span, into a European-dominated international organization with a bureaucracy, a hierarchical, centralized structure and a headquarters based in Amsterdam. By social movement organization standards, this organization was, and still is, (relatively) professional and professionally managed. It has also over much of its life span had at its disposal relatively ample financial resources, not least thanks to its brand name which has helped it to fundraise successfully.

This organization continued its anti-whaling protests, which in the 1970s had appealed to many people and been a media success. However, when it started to campaign against Norwegian whaling, this met with reactions in Norway that were very different from what many Greenpeacers may have expected. These reactions can only be understood against the background of Norwegian history.

A Very Short History of Norway

On the one hand, Norway, with its population of about five million, is a very homogeneous society, both socially and culturally (Hveem et al. 1984). Norway was what Kramer (1984: 92) calls an 'underdeveloped colony' (my translation), first of Denmark and then of Sweden, for more than five hundred years until 1905 (see also Barnes 1954). The society of this 'colony' included many land-owning small farmers and a small upper class of civil servants but had no tradition of nobility, which provided the country with an 'almost unalterable egalitarian social foundation' (Strømsnes, Selle and Grendstad 2009: 399; see also Hansen and Holt-Jensen 1982; Barnes 1954).

Eventually, determined efforts were made in Norway to both define a distinctly Norwegian culture (Strømsnes, Selle and Grendstad 2009; Barnes 1954) and gain national independence, and this provided an incentive for Norwegians to overlook their regional differences (Kramer 1984).

Norwegians rejoiced in the national independence they finally gained in 1905. To this day they continue to be friendly towards their nation-state and its institutions: Grendstad et al. (2006: 133f.) report trust rates of between 50 and 80 per cent in most Norwegian institutions, with a trust rate of 77 per cent in both the legal and the national political system (cf. Sørhaug 1984). The Norwegian political system is perceived by many as an ideal, and there is a strong sense of solidarity and cohesion among Norwegians (Hveem et al. 1984).

My Norwegian interview partners also emphasize the fact that Norwegians have a strong dislike of foreigners 'coming up here and telling them what to do' (see also Nordeng 1984). In fact, the most well-known Norwegian folk tales about *Askeladen* (The Ash Lad –comparable to Cinderella), which are often understood to refer to the Norwegian people, are about a young boy who is weaker than his older brothers and gets bullied by them. But because he is cleverer and more warm-hearted than they are, he triumphs in the end. The Nazi occupation of Norway during the Second World War, unsurprisingly, did nothing to weaken these national feelings of Norwegians.

On the other hand, Norway is not only a homogeneous, but also a diverse society, in the sense that it has what Strømsnes, Selle and Grendstad (2009) and Grendstad et al. (2006) call the 'local community perspective'. For centuries, there were practically as many local cultures in Norway as there were local settlements, given the geographical isolation of most of them because of mountains and fjords, and the relative freedom, self-confidence and economic independence of Norwegian farmers vis-à-vis any kind of ruling class (Hansen and Holt-Jensen 1982; cf. Sørhaug 1984). (Norway is also a diverse society in the sense that there are various ethnic groups, including for example the Sami and Kvener, as well as immigrants, who live there.)

Of course, when speaking of such historical freedom and independence of Norwegians, it must immediately be added that during the 'colonial' time – and up until after the Second World War – Norway was rather poor. Only a small proportion of Norwegian territory is arable land; thus, many farmers were also fishermen, hunters, whalers and sealers, and had to be to survive. This meant that the local cultures were directly tied to the local resources for sustenance (Hansen and Holt-Jensen 1982). From the early 1900s to the 1950s, the Vestfold region of Norway, southwest of Oslo, was the human centre of the increasingly globalizing whaling industry. Most European harpooners and many crewmen came from there. Their highly paid jobs were important to the entire Norwegian economy (Dorsey 2013: 23–27; see also Norsk Kulturråd 2011).

In Norway, nature was and is seen as an environment where people make a living (Strømsnes, Selle and Grendstad 2009; Grendstad et al. 2006). Even today, older generations in particular see fish and whale meat as a regular part of their diet. Generally speaking, the average Norwegian does not consider using the environment, killing animals and so forth in any way a violation of nature (Norsk Gallup Institutt AS 2002; Strømsnes, Selle and Grendstad 2009; Grendstad et al. 2006). As my British interviewee John puts it:

> *John*: Norwegians have a very different attitude towards nature, in general, compared to British people. Nature is a resource to be exploited. If you're in England, you got so little nature, you protect it at any cost. You have a sort of Disney-fied view of nature. Bambi getting shot is horrific to you. Norwegians think that real environmentalism is to go out hunting. Real environmentalism is to go out into nature and be in 'the nature'.

In 2002, the Nature and Environmental Barometer of Norwegian Gallup Institute (an opinion research centre) found that more than 60 per cent of Norwegians were generally positive towards hunting. (Men were more positive than women, and the inhabitants of the Oslo/Akershus region were least positive.) Almost 80 per cent of respondents said that they had good faith in the claim that hunting in Norway was done in a humane way. (Again, city dwellers had least faith here.) Around 80 per cent of respondents said it was important to them to have good walking and hiking grounds reasonably close to where they live and 90 per cent thought more school teaching should take place outdoors. The '*allemannsretten*' ('Every man's right') had strong support. This is a special Scandinavian legal tradition, which maintains that everybody is allowed to enjoy nature and reap her fruits without asking the legal owner for permission, as long as the owner or other people are not put at a disadvantage by that. This could mean, for example, hiking through grounds which are privately owned.

Norwegian society can seem somewhat surprising to non-Norwegians in that, while national feelings in the sense explained above are strong, the small local community, the periphery, has traditionally been seen as the heart of Norwegian identity and culture (Grendstad et al. 2006). Local governments are politically strong (Hveem et al. 1984; Strømsnes, Selle and Grendstad 2009; Grendstad et al. 2006), and Norwegians' wish to keep it that way is probably an important reason for their refusal to join the European Union (Hveem et al. 1984). Even today, it is important to most Norwegians to emphasize their bond with the local community they come from, with its own local dialect, traditional clothing (*bunad*) and so on (see also Kramer 1984).

It is therefore often seen as a problem that the population in the more remote parts of the country has been diminishing for decades. The idea that a decentralized settlement structure should be maintained is strong. Since the 1960s, official government policies have been in place to strengthen the periphery (Hansen and Holt-Jensen 1982; see also Norwegian Ministry of Foreign Affairs 2003).

One reason for the rural exodus is that incomes from fisheries have declined steadily (see for example Christensen 2014a, 2014b). Norwegian whalers are, in fact, fishermen; contemporary Norwegian whaling is done from small fishing boats with crews of a few men, which go out for fish for the best part of the year and for whales for a few weeks per year (Øen 1993; cf. Whale and Dolphin Conservation Society n.d.). Many Norwegians feel that support for the continuation of whaling is merely a logical implication of their support for a decentralized settlement structure, for small peripheral communities.

Statistics Norway (2010) reports that there were around fifty whaling boats involved in the Norwegian whale hunt from 1984 to 1987, and around thirty from 1993 to 2007. Norwegian whalers have only hunted *småhval*, small or minke whales, since the 1970s (Statistics Norway 2006). Yearly catches of minke whales went steadily down, from 4,035 in 1960, to 3,131 in 1970, to 2,054 in 1980, to below one thousand from 1984 up until recently (Statistics Norway 2010). Although catch quotas have recently been raised to above one thousand, whalers find it difficult to exhaust them (Doyle 2009; Fouché 2008).

It is unlikely that this hunt threatens the existence of the minke whale population in question. The common minke whale is not viewed as threatened, according to the IUCN (International Union for Conservation of Nature) Red List of Threatened Species (www.iucnredlist.org). However, it can be argued, as many Greenpeacers indeed do, that any legal whaling also encourages illegal whaling, as well as lobbying for greater catch quotas and for encroachment on other (more endangered) whale species (see for example Sollied 1994). In other words, as long as whaling is legal, there is always a danger that the industry will again begin to overexploit whale stocks (see for example Wallace 1992; Brøyn 1994; Hesstvedt 1993; Pleym 1999a; for scientific models of the dynamics of overexploitation, see for example Gordon 1954; Clark 2005).

There is a varied and interesting history of environmentalism and environmental protest in Norway. According to Berntsen (2011), civil disobedience was used for the cause of environmental protection in Norway for the first time in 1970 during the Mardøla action, a protest movement against the development of waterfalls for energy production. Demonstrations against the development of the Alta River took place in 1979 and 1981.

The Norwegian environmental organization which is maybe most similar to Greenpeace in that it is issue-oriented, nondemocratic and action-oriented, Bellona, was started in 1986. Bellona may have preempted Greenpeace Norway's potential for popular support to a certain extent, given that it was founded two years before Greenpeace Norway (Grendstad et al. 2006). Bellona supports Norwegian whaling (Kristjánsson 2015), as does Norges Naturvernforbund (The Norwegian Society for the Conservation of Nature) (Aasjord 1993) and Natur og Ungdom (Nature and Youth) (Sørensen 1993). WWF and the Norwegian animal rights organization NOAH, like Greenpeace, oppose Norwegian whaling (Bailey 2009). NOAH is not a core environmental organization in Norway, at least partly due to the fact that in Norway, animal rights are not generally perceived to be a necessary part of environmentalism (Grendstad et al. 2006). While WWF, like Greenpeace, is a strong international player in the field of environmentalism, it does not specialize in direct actions. That Greenpeace is a comparatively strong, international organization which opposed Norwegian whaling with confrontational methods explains why it got into more trouble in Norway compared to the other environmental organizations that were and are active there.

Norway got its first environment minister in 1972 (according to Christensen 2014a, he was also the world's first environment minister). Grendstad et al. (2006: 39) interpret this fact as a sign that the environmental field was coopted early by the Norwegian state. None other than Gro Harlem Brundtland, who as Norway's Prime Minister would strongly oppose Greenpeace on the whaling issue (see below), became Norway's environment minister in 1974. Brundtland, Norwegian Prime Minister in 1981, in 1986–1989 and in 1990–1996, is a Social Democrat who from 1983 to 1987 served as Chairwoman of the United Nations' World Commission on Environment and Development, better known as the Brundtland Commission. The Commission published the highly influential report 'Our Common Future' on sustainable development, commonly known as the Brundtland report.

Greenpeace Comes to Norway: a Clash of Self-descriptions

The first Greenpeace anti-whaling actions confronting Russian whalers had taken place in 1975, 1976 and 1977. In 1978, the Greenpeace ship *Rainbow*

Warrior visited Bergen, Norway, to demand a ban on whaling (Klokk 1978). Greenpeace organized for U.S.-Americans to send postcards protesting Norwegian sealing to NRK (the Norwegian Broadcasting Corporation) and the Norwegian daily *VG* in 1981 and 1982 (Svarstad 1982), and carried out a direct action against sealing in 1983 (Holm 1983). Also in 1983, Greenpeace organized a boycott against Norwegian fisheries products as a protest against Norwegian whaling (Andersen 1983). In the years that followed, Greenpeace and other organizations protesting whaling several times considered using a boycott against Norway as a protest instrument (Kristiansen 1984; Granaasen 1985; Harbo 1985). In 1984, German and Danish Greenpeace activists protested with a direct action against chemical discharge in Fredrikstad, Norway (N. N. 1984).

In 1986, Greenpeace drove a rubber dinghy with a fake whale in tow through Bergen harbour during a folk festival (Gjerde 1986). Later that year, the Greenpeace vessel *Moby Dick* and her rubber dinghies tried to block the Norwegian whaler *Svolværing* outside Båtsfjord in Northern Norway (N. N. 1986b). The Norwegian coastguard boarded the *Moby Dick* and escorted her to Vardø island.

A couple of weeks later, at Vardø harbour, two Greenpeacers boarded the whaler *Nybræla* and climbed up into her crow's nest. They refused to come down again, and the *Nybræla* went out to sea with the two on board. After the journey, the activists agreed to climb down voluntarily and were dropped off again at Vardø (N. N. 1986c). They and several members of the *Moby Dick* crew were fined (N. N. 1986c, 1986d). The *Moby Dick* also went to Oslo to raise awareness of the opposition to whaling (N. N. 1986e).

Also in 1986, a moratorium on commercial whaling, decided by the International Whaling Commission in 1982, became effective. Norway had filed a formal objection against the moratorium, meaning the moratorium decision was not binding for the country. However, under international pressure, not least from the U.S. government (Andresen and Skodvin 2008), Norwegian authorities declared in 1986 that Norwegian commercial whaling would be halted from 1988 (Zangl 1999: 215ff.; N. N. 1986a).

Norwegian whalers protested. On the morning of 29 August 1986, seventeen whaling vessels sailed out of Stamsund in the Lofoten islands and lay in line over Vestfjorden (the Western Fjord). When the U.S. aircraft carrier *Nimitz*, scheduled to take part in a NATO manoeuvre, came into view, all seventeen vessels sailed towards it, then withdrew after it had passed their line (Hegna 1986; N. N. 1986a). Protest leader Steinar Bastesen was quoted as saying 'We hope our message will reach the American government', and that the action was directed against 'environmental protection hysteria' (N. N. 1986a).

Despite whalers' protest, Norwegian commercial whaling was halted from 1988 to 1992, although Norway continued whaling for 'scientific purposes' during this period (Statistics Norway 2010).

The Norwegian Greenpeace office was opened in 1988. In 1989, Norway launched a sea mammal research programme, which *Aftenposten* called 'without a doubt the most extensive Norwegian ocean research program ever', and which was led by Professor Lars Walløe of Oslo University (Mathismoen 1992a, my translation). The programme came to the conclusion that there were about 86,000 minke whales in the North East Atlantic – a result that was accepted by the IWC – and that the stock was large enough to be harvested (Mathismoen 1992a; see also Krog and Talsnes 1992).

According to Grytås (2014), Gro Harlem Brundtland intensified the Norwegian diplomatic activities about whaling from 1991. Quoting an unpublished thesis by Hanne Kristoffersen and an interview with Thorvald Stoltenberg, Grytås writes that in the spring of 1991, Brundtland, Norwegian foreign minister Thorvald Stoltenberg and Lars Walløe met U.S. President George H.W. Bush and U.S. Secretary of State James Baker. During this meeting they got 'strong signals' that there would not be U.S. trade sanctions against Norway because of whaling (p. 334).

In 1992, it was announced that Norwegian commercial whaling would be resumed from 1993. Brundtland pointed out that by filing a reservation against the IWC moratorium in 1982, Norway had reserved the right to resume whaling the moment researchers found that the minke whale stock could sustain harvesting: 'This moment has now come' (Krog and Talsnes 1992, my translation).

Norges Naturvernforbund (The Norwegian Society for the Conservation of Nature) supported the decision, saying that renewable resources can be harvested if this is sustainable from an ecological perspective (Aasjord 1993). Natur og Ungdom (Nature and Youth) also supported the decision (Sørensen 1993). WWF opposed the decision, saying that other countries might also take up commercial whaling again and that stocks could quickly be harvested to extinction when economic interests got too powerful; that stopping whaling need not destroy local communities; and that the decision would hurt Norway's reputation and international environmental work (Hvoslef 1993).

At this point, with Greenpeace established as an organizational actor 'on the ground' in Norway, Norwegian authorities prepared to withstand international protest (Krog and Talsnes 1992; Engesland et al. 1992) and Norwegian whalers cheering that 'this is the greatest thing that happened to us since the Germans were driven out in '45' (Krog and Talsnes 1992, my translation), the 'whale war' between Greenpeace and Norwegians began in earnest. This will be described in Chapter 3.

When Greenpeacers protested against Norwegian whaling, their self-description clashed with the dominant self-description of Norwegians. Both self-descriptions were intimately connected to a certain view of nature and the environment. Because they were diametrically opposed to each other, Norwegians often reacted with incomprehension and bafflement to the Greenpeace protests. And because both Greenpeacers and Norwegians were strongly emotionally attached to their respective self-descriptions, the confrontations between the two groups became highly emotionally charged.

Greenpeacers saw themselves as small but clever Davids against powerful and destructive Goliaths. But the Norwegian whalers, who from the Greenpeace point of view had the role of destructive Goliath in this picture, saw themselves as David. As Norwegians, they were small but clever Askeladen, oppressed for a long time by bigger nations. An important part of this Askeladen self-description was the cultural emphasis on small but relatively independent and self-sustaining local communities. Whaling was part of the foundation of this culture, because for small local communities it was both part of making a living and a way of life. It was something that tied the whalers to the places where they lived, something they had grown up and been socialized with (Kalland 2012: 167f.). To portray them as criminal greedy Goliaths (cf. Mathismoen and Jonassen 1993; Myklebust 1993) who committed 'massacres' and 'barbaric environmental devastation' (Andenæs 1987, my translation) was unfair, the whalers felt, and showed that the protesters did not understand them (see also Kalland 2012: 66–73; Barrett 1993; Maddox 1992). Most Norwegians supported the whalers in this. They, too, did not like foreigners 'coming up here and telling them what to do'. They, too, interpreted the attacks on whaling as attacks on small local communities, and they had a sense of solidarity with these communities and felt that they were what Norway was really about (see also Strømsnes, Selle and Grendstad 2009: 401). And they knew that whalers were fishermen, normal people, fathers and grandfathers in small fishing boats, rather than powerful, destructive, cold-hearted Goliaths on factory ships. My Norwegian interviewee Hans put it thus:

> *Juliane*: Why are Norwegians so fierce about continuing whaling on this small scale…?
> *Hans*: It is almost impossible to explain it without first understanding how Norway… functions. Norway is a country where local communities and the whole question of decentralized living are very, very important and have an enormous priority in national politics. … (*points at a poster*

> *with a picture of the Lofoten islands*) Look at the picture, people live up there. Today Norway is one of the richest countries in the world. We have not been that for a long time. Fifty years ago, people up here were very, very, very poor. And they had to live off everything they could. They lived off fish one part of the year. Farming another part of the year and whaling a third part of the year.
>
> So ... you can just see, Germans coming up here. They have money and they are professional activists. They have no clue, they have never seen a live animal being slaughtered. And they just come up here to protest against the whaling. Can you imagine how people in Norway would feel? They are so incredibly provoked by this. Because, I mean, what the hell do you know about this question! What do you know about the history!
>
> So ... it's such a clash of cultures. And even though the majority of the people in Norway don't give a shit about the whaling, pro or against, they would have a lot of sympathy for the people in the area. And for the history of the people in this area. And Greenpeace never, never understood this.

The self-description of Norwegians as small David with an independent spirit was inextricably tied to the Norwegian view of nature as an environment where people make a living. While this perspective might not have been identical to the holistic, biocentric ecology perspective that was dominant in Greenpeace, it would have been relatively compatible with it insofar as it required the human use of nature to be sustainable. But Greenpeacers also saw whales as supreme beings and symbols for environmental survival. Most Norwegians, and particularly Norwegian whalers, could simply not understand this (see for example Barrett 1993). The idea that an animal is beautiful, intelligent or peaceful was simply not an argument to them that you should not kill and eat it, as long as the population of animals is not threatened. And they did not believe that this was the case, because the Norwegian government was controlling whaling and setting catch quotas, and Norwegian scientists were quoted in newspapers as saying that the Norwegian minke whale hunt was sustainable, and Norwegians trusted both groups (see for example Waagbø 1992; Cramb 1994). Most Norwegians did not understand the outrage about whaling; pictures of whales being killed looked normal and familiar to them. Thus, the Greenpeace actions against whaling did not convey the Greenpeace self-description as 'nonviolent protectors of the lives of others' to Norwegians, either. As far as most Norwegians were concerned, there were no precious and endangered lives involved which Greenpeacers could meaningfully have protected (see for example N. N. 1992). Rather, Greenpeacers were using force to prevent whalers from making *their* living (see for example Jonassen 1993; Wallace 1992).

> *Hans:* Average Norwegians didn't see what this symbol represented for Greenpeace. They just saw the opposition to whaling, which they just didn't understand, because is whaling any worse, than how we treat animals all over the world? – No. Is it worse than bull fighting? – No. Is it worse than fox hunting? – No. (See also Maddox 1992)

It should be noted that Norwegian opinion and Norwegian media reporting on the whaling issue were never completely uniform. Thus, in 1983, an article in the Norwegian daily *Aftenposten* began with the sentence: 'I am afraid that many Norwegians have become so occupied with Greenpeace's way of anti-whaling campaigning that we forget what this is really about'. The article argued that it is difficult to control a moratorium on the harvest of endangered whale species if other species are still being legally hunted. Also, Norway's attitude towards the whaling issue made it easier for the Soviet Union and Japan to continue whaling, which could end with the extinction of some species. 'Can Norway, as one of the world's richest countries, not absorb the decline in employment and income which will occur for a relatively small group … in our economy?' (Ruud 1983, my translation). In 1984, two whalers held a press conference in Oslo together with Greenpeace and several other environmental organizations, demanding that the authorities provide better compensation. Whaling quotas were too low for the hunt to be economically viable for them, yet they were not eligible for unemployment benefits as long as they kept their vessels. The whalers wanted to remain anonymous 'for fear of being mobbed by colleagues', and underlined their differences with Greenpeace (Hesstvedt 1984, my translation). Yet this shows that dialogue between Greenpeace and whalers was possible at various points in time.

Thor Heyerdahl, 'the great son of Norway' who was extremely popular in his native country, spoke out against Norwegian whaling in 1985, arguing that whales should not be killed with harpoons and that Norway was ruining its international reputation by continuing whaling. Heyerdahl said Greenpeacers were good folks with good ideals: 'Even if they sometimes go too far, they are better than all those who simply sit with their hands in their laps and let things go over the edge' (Strand 1985, my translation). A 1986 *Aftenposten* article criticized Gro Harlem Brundtland and Norway for insisting that whaling was part of their culture and tradition, calling whaling 'one of the greatest blemishes in the history of our species' (Rasmussen 1986, my translation). Even in the 1990s, national dailies printed articles arguing that although the Norwegian minke whale hunt was sustainable, it was stupid not to cooperate with one's international partners. Bigger whaling nations might follow Norway's example, and Norway's international credibility and reputation would suffer (Yngland 1992; Mathismoen 1992a; see also Chapter 3).

Even Norwegian policy was split for a relatively long time. In 1972, Norway voted in favour of a (nonbinding) resolution that included a commercial whaling moratorium at the UN Conference on the Human Environment in Stockholm, and against such a moratorium at the IWC meeting in London. When an environmental NGO requested clarification of Oslo's position, this 'set off a tiff within Norway's government' between the Ministry of the Environment and the Foreign Ministry (Dorsey 2013: 225). From 1972 until 1982 (the year the IWC moratorium was passed), Norwegian government officials stated that a whaling moratorium had no scientific basis and that a ban on catching minke whales would lead to the depopulation of Northern Norway. At the same time, the Ministry of the Environment was not satisfied with the reasons for opposing the moratorium and managed to get Norway to abstain on the issue on a few occasions. There were concerns that Norway would be hurt by international sanctions and boycotts. In 1982, 'Norway finally had to come to grips' with the inconsistency of its whaling policy, and filed its objection to the moratorium (Dorsey 2013: 268–270). But even after this, the disagreements about the whaling issue within the Norwegian government did not end (Grytås 2014: 333–334, who quotes an unpublished thesis by Hanne Kristoffersen and an interview with a former Norwegian Minister of Fisheries; Tamnes 1997: 319).

In short, it might have been possible at various points in time for Greenpeace and other environmental organizations to convince at least a part of the Norwegian population that whaling was the wrong thing to do. However, the Norwegian government of the early 1990s was prepared to take on the whaling fight. Norwegian whalers and their allies organized and mobilized. And Greenpeace's campaigning against whaling in many instances turned out to be counterproductive, making the whale campaign all but unwinnable in Norway and locking Greenpeace Norway into a situation where it could not succeed, no matter what it did. To understand these dynamics, and to understand how Greenpeace's situation was eventually partly remedied in the late 1990s, it is necessary to know a little bit about the history of Greenpeace in Sweden and Denmark.

Notes

1. Parts of chapters 1, 3, 4 and 5 were earlier published in Riese, J. 2014. 'On Dynamic Processes of Framing, Counterframing, and Reframing: The Case of the Greenpeace Whale Campaign in Norway', in: K. Fahlenbrach, E. Sivertsen, and R. Werenskjold (eds), *Media and Revolt. Strategies and Performances from the 1960s to the Present.* New York/ Oxford: Berghahn Books, pp. 283–299. Republished with permission.
2. Dale (1996: 211) notes that rather than being the originator of the term 'global village', McLuhan seems to have borrowed the concept from Wyndham Lewis, but Eric McLuhan (n.d.) thinks that his father was in fact the originator of the term.
3. Gregory Bateson, whose work this book draws heavily on, was one of the researchers involved in this; see Burnett 2012; Bateson 1983a.
4. Zelko quotes the following source here: S.D. Cassidy, 1992. *Mind Bombs and Whale Songs: Greenpeace and the News.* PhD diss., University of Oregon.
5. In the framework of autopoietic systems theory, the concept of self-description is partly equivalent to what is generally known as organizational identity in organization studies (cf. Seidl 2003; on the concept of organizational identity, see for example Reger, Gustafson, Demarie and Mullane 1994; Dutton and Dukerich 2004; Nag, Corley and Gioia 2007).

Chapter 2

'Greenpeace Should Be a People Persuader and Stand United Internationally'

Greenpeace in Sweden and Denmark

The following Greenpeacers are quoted in this book:

From the Stockholm Greenpeace office:

Lars	Executive Director, Greenpeace Sweden and later Greenpeace Nordic, 1997 to 2002
Nils	Volunteer and activist, Greenpeace Sweden, 1985 to 1987 Action coordinator, Greenpeace Sweden, 1988 to 1990 Campaigner, Greenpeace Sweden, 1990 to 1994 Executive Director and Campaign Director, Greenpeace Denmark, 1994 to 1996; in addition, stood in as Fundraising Director for Greenpeace Denmark for a period of time Campaign Director, Greenpeace Sweden and later Greenpeace Nordic, 1996 to 1999
John	Fundraising Director, Greenpeace UK, 1992 to 1995 Fundraising Director, Greenpeace Sweden and later Greenpeace Nordic, 1995 to 2000
Moa	Director of Finances, Greenpeace Sweden and later Greenpeace Nordic, 1995 to 1999
Caroline	Campaigner
Alex	Campaigner
Björn	Press Officer
Astrid	Fundraiser

From the Oslo Greenpeace office:

Hans	Campaigner
Søren	Campaigner
Mari	Campaigner
Henrik	Campaigner
Sander	Campaigner
Tor	Campaigner
Fredrik	Campaigner
Mikael	Activist

Greenpeace Sweden: the Early Years

The story of Greenpeace Sweden, founded in Gothenburg in 1983 (Erwood 2011: 40), introduces some people who will be key figures in this book and an important Greenpeace campaign philosophy.

According to Nils, a Swede who worked for Greenpeace Sweden as a volunteer from 1985 to 1987 and as an employee from 1988 to 1994, during its first years Greenpeace Sweden experienced 'gigantic growth' and 'a great lot of sympathy'. Greenpeace Sweden's supporter figures show 158,030 supporters in 1987 and 210,000 in 1991. In Sweden, like in other countries such as Germany, the dangers of environmental destruction received much public attention in the 1980s; the Swedish Green Party (miljöpartiet) was founded in 1981. The 1985 bombing of the Greenpeace ship *Rainbow Warrior* in Auckland, New Zealand by the French secret service, which claimed the life of photographer Fernando Pereira (King 1986; Erwood 2011: 58ff.), drew additional attention and support to Greenpeace. According to Moa, at one point Greenpeace Sweden had forty-five staff in a more than 1,000-square metre office in Gothenburg.

Indeed, there had been public interest in environmental questions in Sweden for quite a long time before the first Greenpeace office was opened there. Jamison, Eyerman, Cramer and Læssøe (1990) agree with my interviewees that Greenpeace was remarkably successful in Sweden (until the time of their book's publication in 1990). According to these authors, the context of this was that there had long been a strong state interest in environmental issues in Sweden, and such issues quickly became parliamentarized. According to Jamison et al., Swedish environmentalism was often a product of dialogue with and within the establishment. 'Utopian' elements of environmentalism gained comparatively little foothold in Sweden. Jamison et al. interpret Greenpeace's success as a result of the organization's fit with this Swedish political context and culture.

What my interviewees tell me about Greenpeace Sweden in parts matches Jamison et al.'s 1990 assessment. Alex, who started working for Greenpeace in 1989 and for Greenpeace Sweden in 1994, says that 'in the year and a half since I started working for Greenpeace Sweden we didn't do a single action or a single demo. Nothing. You know, we were like, (*mockingly*) writing reports and writing like, official replies…'. At the same time, in Alex's view, Greenpeace Sweden started too many projects at once, and was brought in on political processes by the government, being included and involved in finding compromises and not taking a sufficiently activist and radical approach: 'You know, we're presenting ourselves like, the cool heroes for the environment in inflatables, whereas in reality we're a bunch of fucking bureaucrats who're afraid to say anything'. And John (who joined Greenpeace Sweden in 1995) chimes in: 'All these people who joined the organization were looking to ride on this lovely wave, everyone loved them, people would buy them drinks in the bar…'.

According to my interviewees, in its early years Greenpeace Sweden was not a sufficiently professional and focused organization, either. This applied to both campaigning and fundraising. Through unsophisticated advertising and, in John's words, 'sending out rotten useless advertising and mail packs', Greenpeace Sweden was able, for a while, to get huge numbers of people to become supporters and to get huge fundraising returns. Nils claims that Greenpeace Sweden hardly paid attention to getting people to renew their membership every year; instead it just recruited new supporters.

At the same time that this way of running the organization turned out to be unsustainable, from the late 1980s, Magnus Gudmundsson, an Icelandic film maker, published several TV documentaries. The first of these was 'Survival in the High North' of 1989, which was shown in all Nordic countries (Rasmussen 1989). Gudmundsson claimed at different points in time that Greenpeace had used false footage in its documentaries, that it controlled secret bank accounts and that it cooperated with groups that had not renounced violence. Greenpeace successfully sued Gudmundsson in Norway over some of his charges, but the media attention was huge in any case (Motavalli 1995; Bentzrød 1992a; Dale 1996: 28f.; see also Olsen 1994).

My interviewees say that during its first years Greenpeace Sweden received lots of positive media reports, and that the Gudmundsson films were part of the second wave of coverage which was much more critical. John interprets this as 'raise them up first and then you chop them down, standard tabloidish journalism' which ensures circulation; Nils calls it a 'media backlash'. And Greenpeace Sweden, which was used to being praised, did not have the know-how necessary to cope with this kind of media dynamics. It had no idea how to react to the Gudmundsson films. According to John, 'They reacted with

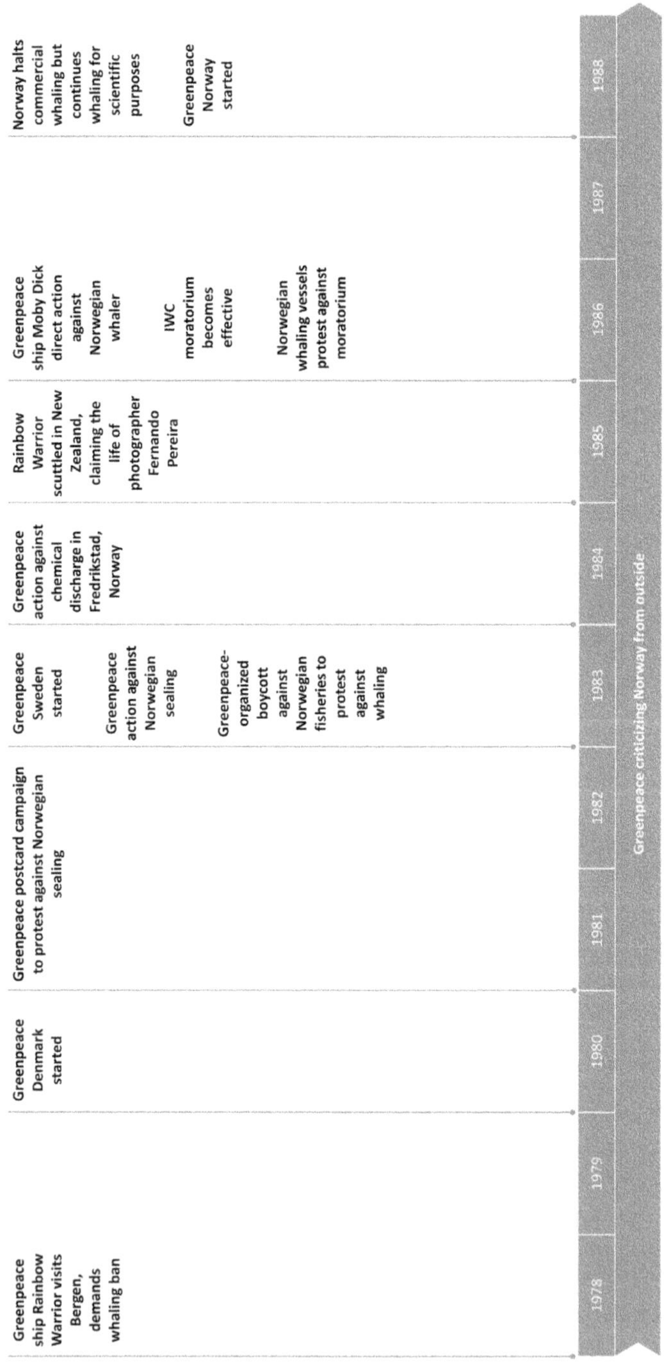
Greenpeace ship Rainbow Warrior visits Bergen, demands whaling ban
Greenpeace Denmark started
Greenpeace postcard campaign to protest against Norwegian sealing
Greenpeace Sweden started
Greenpeace action against Norwegian sealing
Greenpeace-organized boycott against Norwegian fisheries to protest against whaling
Greenpeace action against chemical discharge in Fredrikstad, Norway
Rainbow Warrior scuttled in New Zealand, claiming the life of photographer Fernando Pereira
Greenpeace ship Moby Dick direct action against Norwegian whaler
IWC moratorium becomes effective
Norwegian whaling vessels protest against moratorium
Norway halts commercial whaling but continues whaling for scientific purposes
Greenpeace Norway started
1978
1979
1980
1981
1982
1983
1984
1985
1986
1987
1988
Greenpeace criticizing Norway from outside

Figure 2.1. Timeline of events discussed in this book

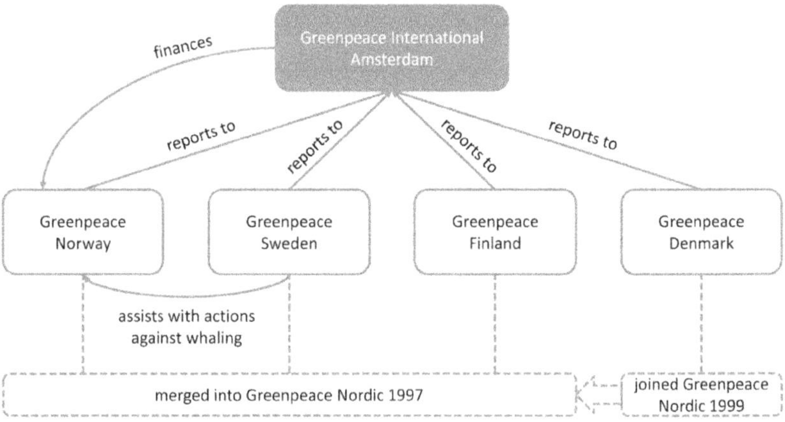

Figure 2.2. The Nordic Greenpeace offices before 1997

total panic. They were wondering whether it was true. You had people in the organization who weren't sure about what was right and what was wrong. They sort of felt, this is wrong, but maybe that bit's right'.

Whether or not the criticisms directed at Greenpeace in the Gudmundsson films (and other critical media coverage) were justified, according to my interviewees what they did reveal were Greenpeace Sweden's lack of professionalism and lack of focus in campaigning and media work. Or as Alex puts it:

> The Gudmundsson movie was like, a boy crying the king is naked, and even though the king actually is clothed, and the boy is lying, you start having a closer look at the king, you go: uh-oh! (*Laughs*) OK, maybe he is not naked, but boy! Something's seriously wrong with the king.

Greenpeace Sweden's supporter figures peaked in 1991 (210,000), and then declined continuously: 160,000 supporters in 1992; 154,000 supporters in 1993; 119,000 in 1994. (Greenpeace accepts money from individuals only, not from corporations etc.) The Gudmundsson films were certainly not the only reason for this, although they were probably important (N. N. 1995c). My Swedish interviewee Björn points out that there may have been a general decline in public interest with regard to environmental questions during that time. Whatever the reasons were, the reduction in income induced 'turmoil' in the organization (Alex). According to John, 'a lot of people who'd had a life so easy stopped. They were not in it for the long term really, they were not the sort of people who the organization needed anyway'. After a tumultuous period, Greenpeace Sweden moved from Gothenburg to the Swedish capital Stockholm in 1995 (the fundraising department remained in Gothenburg until 1996). It lost almost all its staff, and started to recruit new staff and rebuild the organization. According to Moa, Greenpeace Sweden had eighteen permanent employees in a 300-square metre office after the relocation. The early years of Greenpeace Sweden, in which, in Alex's words, the organization had been 'too big and fat and lazy and not doing what we should be doing, you know', were over.

The Ultimate Campaigning Organization

At this point, John and Nils, who would subsequently become change agents with respect to the campaign against Norwegian whaling, took leading positions in Greenpeace Sweden. John, who is British, was the Greenpeace U.K. fundraising director at the time and did some research and advisory work for Greenpeace Sweden's fundraising in 1995, which later led to him taking on the job as Greenpeace Sweden's fundraising director. By his own account, his first job after university had been to sell cable TV. Then one day, 'Bob Geldof swore at me on TV and said, you better get off your ass and do something. … I thought he was talking to me. So I thought, okay, this means I have to do something'. John says he felt guilty about the fact that so much in the world was not right and about his own privileged position, and seeing Geldof on TV seemed like a message that merely donating some money to charity would not be enough for the guilt to stop. He had to do more than that. John resigned from his job the next day and applied 'for a whole series of jobs. And ended up at a global development agency called ActionAid'. He did market research at first, subsequently held different jobs in the organization, and in 1992 got the job as fundraising director for Greenpeace U.K. in London.

John would work for Greenpeace for ten years in total (with a two-year hiatus in between), in different positions, and would come to be highly respected by his Greenpeace colleagues. My other interviewees see him as a very capable fundraiser, but above all they admire his analytical and emotional intelligence and his ability to support those who reported to him, so that they could develop their talents and flourish in their jobs. He was demanding and uncompromising with respect to his idea that Greenpeace must follow its ideals and reach its goals, and never lose sight of those.

> *Alex:* When John first came to Greenpeace Sweden, he kind of made a round, sat down and talked with everybody. And he talked to me, he said: 'So, why are you doing what you're doing?' I was like: 'What?' He was like: 'What are you doing right now?' I was like: 'Well, I'm doing

> some research on uranium'. 'Oh, why're you doing that?' 'I'm trying to put together a general picture of the adverse effects of uranium'. 'Yeah but why're you doing that?' I was like: 'Because I want to expose the fact that supposedly clean nuclear power is actually paid for by dirty environment…'. 'Yeah but why're you doing that?' 'Well, because I want to use that to make political gains as well as public gains so that we can come closer to phasing out nuclear power…'. 'But why? Why're you doing that?'
>
> He just kept pushing. I was like: 'What're you trying to get at?' And he said: 'I'll tell you why you're doing that. You're doing that because you care about the planet and you want to save it'. I was like: 'Well yeah'.
>
> *Juliane and Alex both laugh*
>
> *Alex:* This was great, you know. John is lovely.

John's demanding and insistent goal-orientedness is also illustrated by his reaction to my interview question as to why he, as a fundraising director, some years later invested a lot of his time and energy in the campaign against Norwegian whaling. At first, John misinterpreted the question, answering that he did this because he began to understand the problems with the whale campaign (more on this below). I then rephrased my point, saying that he was employed by Greenpeace as a fundraiser, not as a whale campaigner, so it was not obvious that he would or should work hard on the whale campaign, in addition to his other duties. To this, John replied that 'it was about me being an environmentalist and doing this stuff not because I want to have lots of money, but because I want to change the world'.

Yet John would come to feel for Greenpeace what he described to me as 'love-hate', and eventually he would give up on the organization and leave (in 2004). At that point he no longer felt that Greenpeace had as much power to change the world as he wanted it to have. He would go on to work as a consultant and advisor to civil society organizations around the world.

Nils, who is Swedish, became the campaign director of Greenpeace Sweden in 1996. Like John, Nils was demanding of himself and of others. He generally gave an impression of calm and self-confidence. He says that when he was younger, he had joined different Swedish nonprofit organizations that were part of the peace and environmental movements, but soon felt that 'it's a lot of sitting around talking and very little that gets done'.

> So then the first thing I did was, I thought that well, then I have to do something myself. And so I moved to the countryside and began to farm and built a greenhouse and had a bakery. But … it felt like the world just continued to get worse and worse around me however much

> I farmed (*chuckles*) and tried to live as a good example. And I was living on an island outside of Gothenburg back then, and I saw how the sea all around got more and more destroyed. And felt that what I am doing is going too slowly. And then I thought that Greenpeace was an organization which got something done, which changed things. Did things and changed the world.

In addition to farming and running a bakery, Nils had also worked as a family therapist. He says that he learned much about the functioning of groups from this work, for example 'how groups and systems fight against change. And how you can bypass that resistance. That knowledge is useful when you work as a campaigner, as well (*chuckles*)'.

After working for Greenpeace Sweden as a volunteer from 1985 to 1987, Nils was employed by Greenpeace Sweden as action coordinator and campaigner from 1988 to 1994. He does not give himself very much credit for his work during those years. Whatever the quality of his campaign work back then may have been, he must have learned a lot along the way. He became Greenpeace Denmark's Executive Director in 1994 (more on this below), campaign director for Greenpeace Sweden in 1996, and eventually, a well-known advisor and trainer for different Greenpeace offices and other civil society organizations around the world.

> *Nils*: When I started at Greenpeace, there was no one who talked to me about what a campaigner does, how you do good campaigns. I think it's very few within Greenpeace who have had briefing about that. ... There was no campaign theory, like, in the first place. So it was first years later when I was asked to start training other offices within Greenpeace, in campaigning, that I realized that there is like, no methodology.... Then I started to put it together myself.

The fact that Nils was well-connected in the international Greenpeace world came in useful when he and John took charge of the campaign against Norwegian whaling (see Chapter 4). Among my interviewees there is at least as much respect and admiration for Nils as there is for John. Some of those who worked for these two in the second half of the 1990s positively adored them.

> *Alex*: Nils was really a leader. And I think he did a great job. Recreating the Swedish Greenpeace office. Like starting from zero, I think it was excellent. ... Nils comes in, in 1996, and he is the guru building it up. We all look up to him, he is the one to tell us what Greenpeace is, what campaigning is, and we all kind of like go 'fearless leader, we go' and we do it.

Yet Nils was also perceived by some who reported to him as overly demanding. (Nils himself says diplomatically that 'there were quite a lot of internal tensions' in the Greenpeace office.) One begins to understand why when reading internal Greenpeace documents that report on Nils' vision for campaigns, which he tried to develop as campaign director in Sweden: 'The Ultimate Campaign Department', aka '*Phyllis Cormack* Campaigning'. As the latter term indicates, this vision consciously attempts to relate to the original, or in Alex's words the '*ursprüngliche*', thinking of the very early days of Greenpeace (*Phyllis Cormack* was the name of the very first Greenpeace ship, baptized *Greenpeace* for the voyage to Amchitka; Hunter 2011, Bohlen 2001). Nils described this as idealistic, action-oriented, radical ('since industry teams up with rational environmentalists, it is Greenpeace's role to be radical'), highly cooperative both internally as well as with regard to external networks, and team- and project-oriented. In his interview, Lars, who became the Executive Director in 1997, clearly agrees with Nils' vision that Greenpeace must be radical.

The overriding idea was to 'send the staff out of the office' to get trained as activists, undertake actions, communicate. Björn reports that Greenpeace activists sat in a forest they were trying to protect over Christmas, no less, although police and media attention had been minimal for weeks. Their doggedness was rewarded: the forest campaign developed into a media and political success for Greenpeace.[1]

Astrid reports how, when she first started to work for Greenpeace Sweden, the staff sat in small offices, separated from each other. This changed when Nils became the campaign director. The walls were taken down, and all staff, including press and fundraising people, got involved in and focused on campaign work. A number of people would be involved in the planning of a campaign; almost everyone would be involved in the roll-out.

John, the then Fundraising Director, agreed with Nils about this approach to running the campaigns and the office. According to Moa, John clearly communicated that all staff, including fundraising staff, should be part of Greenpeace's campaigning and contribute to it. Everybody should be updated on what campaigns were doing and what was happening, so that they could answer telephone calls properly even if they weren't campaigners.

The ultimate campaigning approach seems to have been successful in boosting both the staff's commitment and their team spirit. Alex says: 'There was a longish process over a year or two, until we were up to strength, but then like, God damn it! – we got a new start, it was like Phoenix being reborn from the ashes'.

The ultimate campaigning approach is designed to realize a vision of Greenpeace which I call the 'Greenpeace should be a People Persuader' vision. For people who take this view, the important thing is not what Greenpeace does, it is what Greenpeace gets the people to do. The goal is to create an atmosphere in society 'whereby the change we want to achieve is not only possible, but unavoidable' (Alex). Greenpeace must win a majority of people over to support Greenpeace's demands, and then legislation etc. will follow. On its own, Greenpeace doesn't mean very much, but if Greenpeace moves the people, change will be unstoppable.

Indeed, John, who strongly supported Nils in building 'the ultimate campaigning organization', goes so far as to say that the highest goal of such an organization might be empowering people rather than winning certain campaigns.

> *John*: I think a lot of stuff that we do is not the things that we're campaigning about. It's about changing the world, the way people feel about their place in society, and what they can do, what they can achieve. The fundamental thing that happened when Brent Spar happened is much bigger than winning that campaign. Beyond that, the whole of Europe changed the way they felt about what they were able to do. It didn't last for very long. But for a period of about a year, the mood of people became that you could change things, you could do things and things might happen if you all got together. (Cf. Carmin and Balser 2002)

The Brent Spar was an oil storage and tanker-loading buoy owned by the Shell oil company. In 1995, Shell announced plans to dispose of it at sea. Greenpeace occupied the Brent Spar, demanding that it be dismantled on land in an ecologically sound way. The action received enormous media attention and triggered much public protest against Shell, including a huge consumer boycott in Germany. Eventually Shell decided not to sink the rig and to dismantle it on land instead. Also in 1995, the OSPAR Conference, the international mechanism responsible for the environmental protection of the Northeast Atlantic, decided by majority vote to generally prohibit the sinking of oil platforms in the North Sea and the Northeast Atlantic (Erwood 2011: 100ff.; see also Rose 1998; Livesey 2001).[2]

Direct actions like the Brent Spar action are essential in persuading and empowering people. Yet Nils and John knew, and tried to convey to everyone in Greenpeace Sweden, that in order to persuade and empower people, direct actions – and indeed whole campaigns – must possess certain qualities. They are not simply an instrument to easily attract public attention. They must be apt to actually stop the practice Greenpeace is criticizing and/or effectively expose it to the world, bear witness to it (cf. Carmin and Balser 2002). They must be credible and moving, convey a clear, simple and urgent message, be sustained over time and demonstrate complete commitment. This is not easily achieved, which is why Nils and John wanted to build a committed, credible, focused 'ultimate campaigning organization' with great staying power that could pull it off.

> *Alex*: Of course it's about PR, 'cause the only way you're gonna save the future of the planet is by PR, that's why we do the actions, to communicate our messages out, but the only reason it's good PR is that it's not about PR. As soon as it becomes about PR, it's no longer good PR.

While Greenpeace, according to the 'People Persuader' vision, wants to move people, it is not looking to be popular itself. (Indeed, John chided former staff of Greenpeace Sweden for riding on a wave of popularity and letting people buy them drinks in bars.) In particular, Greenpeace is not looking to be accepted and respected by governments or the establishment, by those who hold power in society. According to Gamson (1975), protest groups can be successful either in the sense that they are accepted by their antagonists as a valid spokesman for a legitimate set of interests and/or in the sense that their beneficiary gains new advantages. The 'Greenpeace as People Persuader, ultimate campaigning organization' vision holds that Greenpeacers are radicals questioning authority and that the goal is to stop environmental destruction. Gamson's second criterion for success is thus clearly given priority over the first. This is reflected in the old Greenpeace slogan 'We work to become unnecessary'.

> *Alex*: New management was new, but really old, traditional, like, *ursprünglich* Greenpeace thinking. You know, we should always be in the opposition, we should be fighting from the outside, we should not be fighting from the inside. And I think that's what happened in the early years of Greenpeace Sweden, we became too popular, and therefore we became mainstream, brought in to the warmth, and we should've stayed out in the cold and that's where Nils and John put us: back out in the cold.

Nils' and John's approach to developing and running Greenpeace Sweden was in some ways a spiritual successor to the early McLuhan-inspired 'change in mass consciousness' Greenpeace philosophy (see Chapter 1). In others it was reminiscent of other Greenpeace roots such as Quakerism. It may be summed up as a belief that through idealistic actions, radical ideas and strategic use of mass media, Greenpeace convinces people that social change is necessary and possible, so that they take collective action and make it happen. What Nils and John were not really interested in was creating or promoting an 'environmental mythology', like some early Greenpeacers had wanted to do. Nils' and John's way of thinking was quite practical, sober and down-to-earth. (As I will explain in Chapter 4, the one possible exception to this was that John was a self-confessed 'arch whale lover'; however, his analysis of the Greenpeace whale campaign in Norway would turn out anything but 'mystical'.)

The two managers approached their project of building the persuasive 'ultimate campaigning organization' with patience and a long-term perspective. They were aware that it is not easy to make change in society unavoidable, something the early years of Greenpeace Sweden had also proven. Their vision of Greenpeace and their campaign philosophy soon afterwards came to bear on the anti-whaling campaign in Norway.

In a process which took place between 1997 and 1999, Greenpeace Sweden merged with Greenpeace Norway, Greenpeace Finland and Greenpeace Denmark to form Greenpeace Nordic. Some knowledge of the historical background of Greenpeace Denmark, as well as of Greenpeace Sweden, is important for an understanding of the process of organizational reflection on, and strategy change in, the whale campaign in Norway, which happened later.

Greenpeace Denmark: Fighting Greenpeace International

Greenpeace Denmark was founded in 1980 (Erwood 2011: 40) and merged with the other Nordic offices in 1999. The part of Greenpeace Denmark's history which I include here illustrates the conflict potential between national and international interests in the global Greenpeace organization. Moreover, it does so in reference to differing attitudes towards the natural environment. This history of the Danish Greenpeace office later contributed to the understanding of the Norwegian whale campaign problem in the unified Greenpeace Nordic office.

In 1996, the international Greenpeace organization ran a campaign against overfishing in the North Sea. The focus of the campaign was on industrial fishing – 'the "hoovering" up of millions of tonnes of tiny fish in fine nets for processing into fish meal and oil', done mostly by Danish fishing vessels at the time (Brown 1996).

Nils was Greenpeace Denmark's Executive Director at the time. He stood in for both campaign director and fundraising director for certain periods of time as well (the Danish Greenpeace office consisted only of a handful of people). He and his Danish colleagues knew that a campaign focused on industrial fishing would be unpopular and hard to defend in Denmark. The general Danish attitude towards nature is rather unsentimental, and the catch from the industrial fisheries in question was used in the Danish agricultural industry (N. N. 1996a), which had a strong standing in the country (Jamison 2004). Indeed, the *Scotsman* of 24 July 1996 quotes a journalist working at the Danish national daily *Politiken* saying that 'the public's sympathies lay with the fishermen because the activists were exaggerating the significance of the issue'.

> 'It looks as if Greenpeace is shouting against a small environmental problem. Many Danes see other environmental issues as being far more important than industrial fishing', he says.
>
> The Danish Fishermen's Association, which acknowledges that Greenpeace has done some good for the North Sea in the past, agrees that Greenpeace has not picked the right target this time.
>
> Jens Muller, its deputy director, says: 'Every form of fishing has an effect on stocks, but it is our opinion that [the type of fishing targeted by Greenpeace] is only having a minor effect because only a small proportion of the stocks are being taken'. (Dalton 1996)

According to Nils, the problem was that Danish Greenpeacers themselves thought that the campaign was misdirected and had a hard time arguing for it. He insists that they would have been ready to take on a fight running an unpopular campaign which they thought was legitimate. But they thought that to attack Danish industrial fishing in particular, when other nations did other types of fishing in the North Sea which could also be judged to be environmentally destructive, was to take sides in an unfair way.

This clashed with the campaign philosophy of the British Greenpeace office. Nils says that British Greenpeacers wanted to focus on an issue which made ethical questions of right and wrong perfectly clear-cut and could be used as a symbol of the problem of overfishing more generally. 'Hoovering' the bottom of the sea, in some cases taking the last fish there is, to turn it into products such as animal feed seems more obviously ethically wrong to many people (though not necessarily Danes) than fishing for human consumption etc., even though the latter may be just as damaging to stocks.

As a consequence of this fundamental disagreement between the Danish Greenpeacers and the other Greenpeace chapters involved, the Greenpeace campaigner who was responsible for the fisheries campaign in Denmark publicly criticized the campaign (N. N. 1996e). Nils reports that he made it clear to the campaigner that this was unacceptable, and agreed with him that he would resign. However, the Board of Greenpeace Denmark decided against this, and instead asked Nils to issue a media statement saying that Greenpeace Denmark supported the critical campaigner and distanced itself from Greenpeace International.

Nils refused to do this, taking the view that if Greenpeace, as an international organization, had decided to run a campaign, then national chapters had to support that campaign. He was fired from his position of Executive Director on 1 July 1996 (N. N. 1996d; N. N. 1996e). Greenpeace International sent its Executive Director Thilo Bode and his deputy David Newman to Copenhagen for crisis meetings (Randsborg 1996).

> *Nils*: All this fighting became such a media issue in Denmark, front pages… It was really so stupid all of it, GPI reacted of course, Thilo who was the boss then, went up to Denmark and met the Board, I don't know exactly what he said, but he probably said that now you gotta take back all of that, otherwise we'll shut this office down. Moreover, Greenpeace Denmark was economically dependent on GPI, it was quite easy for GPI to say this… so then the Board had to go out and say to the media that we have discussed this and we take back what we said.

On 3 July 1996, the critical campaigner 'was transferred to other duties from his position as leader of Greenpeace's Denmark Fishery campaign [under pressure from Greenpeace's international leader Thilo Bode]' (N. N. 1996e). The Danish Greenpeace chapter had been 'brought to its knees' and 'whipped back into line' by Greenpeace International who had 'shown muscle' (Randsborg 1996, my translation).

Nils became the campaign director of Greenpeace Sweden and later Greenpeace Nordic. According to Alex, the feud about the fisheries campaign contributed to the ongoing decline of the Danish Greenpeace office. This decline was only reversed by the office joining Greenpeace Nordic in 1999. Somewhat ironically, this merger again made Nils the campaign director of Greenpeace Denmark.

To sum up, the Danish office, when it joined Greenpeace Nordic in 1999, already had some experience regarding national versus international interests inside the Greenpeace organization. It also had some experience when it came to cultural differences between countries regarding attitudes towards nature and to conflicts between different campaign philosophies in different Greenpeace offices. And Nils, who had been the Danish and Swedish campaign director and was now the Nordic campaign director, strongly emphasized international strategic and brand thinking and was ready to fight for this kind of thinking inside the organization.

Notes

1. Kihlberg (2011) has interviewed officials from Swedish county governments who are knowledgeable about the cases Greenpeace highlighted in 1997 and 1998. Regarding two of these cases, they are quoted as saying that without Greenpeace's actions the forest areas in question would not have been made reservations. Kihlberg also highlights the media interest in the actions and the fact that the issue became more prominent on the political agenda.
2. Niklas Luhmann published an article about the Brent Spar action in the German newspaper *Frankfurter Allgemeine Zeitung*, discussing the question of whether firms could learn from the public (Luhmann 1995a).

Chapter 3

'Campaigning Against Each Other'

Greenpeace Norway

The Norwegian 'Whale War' of the 1990s

Before 1988, Greenpeace protests against Norwegian whaling were organized from outside Norway.[1] In Chapter 1 I have described the 'metaphysical' view of whales and the idea that the whale campaign could help to bring about an ecological 'consciousness revolution', both of which were rooted in Greenpeace's countercultural history. I have also described the Greenpeace self-description of being 'Davids against Goliaths' and 'peaceful protectors of defenceless Nature against violent threats'. I have explained why these clashed with the perspectives and self-descriptions of many Norwegians.

In what follows, I will first give a brief chronological account of the events related to the Greenpeace campaign against Norwegian whaling from 1988 to 1998. I then present my systems-theoretic analysis of the campaign during that period, which shows that the campaign became deadlocked in what I call 'chaos communication loops': self-reproducing and self-reinforcing unproductive patterns of communication.

In 1988, the same year that Norway halted commercial whaling in line with the International Whaling Commission moratorium, Greenpeace established a chapter in Norway (Erwood 2011: 40). It was no easy job to build the official Greenpeace presence in Norway after the 'actions from outside' had already bewildered or offended many Norwegians. Mari tells me that when she decided to do volunteer work for Greenpeace in the early years of Greenpeace Norway, her mother called her on the phone to warn her she would never get a 'real job' afterwards. (Luckily for Mari, this prophecy did not come true.) However, the great majority of those who worked for Greenpeace Norway during its first decade were very dedicated and committed to the organization and its goals. Having moved on to different jobs after working for Greenpeace, they still hold Greenpeace in high regard. They think that Greenpeace is knowledgeable, fast and efficient, employs good people, achieves its goals, and makes a difference. Indeed, some of them say Greenpeace is better, more effective, than other civil society organizations. One says it is one of the best environmental organizations in the world. They like the fact that Greenpeace is an international rather than merely a national actor. The fact that these individuals stuck it out in Greenpeace and remained loyal to the organization despite all kinds of difficulties, which are the subject of this chapter, is due to this esteem in which they held and still hold it.

One of the first things the Norwegian Greenpeace office needed to do was deal with Magnus Gudmundsson's film 'Survival in the High North' (see Chapter 2). Greenpeace Norway tried, unsuccessfully, to prevent the film being shown on Norwegian TV (Rasmussen 1989; Jentoft 1989). Gudmundsson was later ordered by a Norwegian court to pay Greenpeace 30,000 Norwegian kroner in damages (Bentzrød 1992a; see also Motavalli 1995; Dale 1996: 28f.).

In 1990, the Greenpeace vessel *Beluga* went on a tour along the Norwegian coast to raise awareness of the increasing pollution of the North Sea. NTBtekst (N. N. 1990) quotes a Greenpeace campaigner as saying 'We want to show that Greenpeace is something more than seals and whales' (my translation).

In June 1992, Norwegian government representatives announced that Norway would resume the commercial hunt of minke whales from 1993 (Krog and Talsnes 1992; Engesland et al. 1992). Also in 1992, Bjørn Økern, the Executive Director of Greenpeace Norway, was fired (Bentzrød 1993). An article in the Norwegian national daily *VG* of 26 July 1992 reported that Økern supported the Norwegian commercial minke whale hunt and said that Greenpeace conducted direct actions to make money (Nielsen 1992).

Also in July 1992, the Greenpeace ship *Solo* followed the Norwegian whaling vessel *Nybræna* in the Barents Sea. The *Solo* crew claimed to have saved the life of a whale by driving a rubber boat between the animal and the *Nybræna* (Tømmerås 1992). In August 1992, professor Leif Ryvarden, a natural scientist, resigned from his position as Chairman of the Board of Greenpeace Norway, criticizing the organization's action against Norwegian whaling and saying that the Norwegian minke whale hunt was acceptable: 'It must be possible to harvest a renewable resource'. Ryvarden also said he would remain a supporting member of Greenpeace, however: 'There are no other organizations that do as much for the environment as Greenpeace' (Waagbø 1992, my translation). In December 1992, the organization Sea Shepherd attempted to scuttle the *Nybræna* (Furuly 1993a; Kalland 2012: 83). Sea Shepherd was founded by Paul Watson, a militant environmentalist who was forced out of Greenpeace in 1977 (Zelko 2013: 270ff.)

In May 1993, the Norwegian national daily *Dagens Næringsliv* reported that it was receiving letters from British citizens who were following a call by Greenpeace to write protest letters against whaling to the four most influential Norwegian newspapers (Wiedswang 1993). On 26 May 1993, a group of Greenpeacers chained themselves to three whaling vessels in Skrova in the Lofoten islands. They terminated the action voluntarily in the evening (Nielsen and Stoltz Bertinussen 1993). (The Skrova action will be discussed in more detail below.) On 27 May 1993, Greenpeace started a campaign for a boycott against Norwegian exports to Germany, Italy and the U.K. (Schiefloe 1993).

In June 1993, Bjørn Økern published a book titled *Makt uten ansvar* (Power without Responsibility), a 140-page attack on Greenpeace. Økern claimed that Greenpeace was a gigantic money machine, that its whale campaign was ridiculous and only served to make money for the organization, and that Greenpeace was a brake block for the environmental movement (Bentzrød 1993, my translation). He was supported at the presentation of the book by Leif Ryvarden, former Chairman of the Board of Greenpeace Norway (Omdahl 1993b).

In January 1994, Sea Shepherd attempted to scuttle the whaling vessel *Senet* (Furuly 1994; Kalland 2012: 83). During the spring of the same year, Greenpeace U.K. organized for postcards protesting against Norwegian whaling to be sent to the Norwegian embassy in London (N. N. 1994a).

In July 1994, five Greenpeace activists from the Greenpeace ship *Solo* boarded the *Senet* when the latter was whaling in international waters in the North Sea, and tried unsuccessfully to steal the harpoon cannon from the *Senet*'s deck. The vessel's skipper, Arvid Enghaugen, said that some of the activists were thrown overboard by the whalers while others jumped overboard themselves; the *Solo* crew claimed that the whalers threw all five activists overboard. Enghaugen is quoted as follows: 'They did not manage to stop the hunt with their sailing round the [*Senet*] and with their video filming of everything we did' (my translation). According to Enghaugen the activists were Danes and Germans (Westengen 1994).

A few days later, the Greenpeace ship *Sirius* conducted a direct action against the *Senet* in the North Sea, during which one activist cut the harpoon line between the vessel and a whale which had been harpooned. Two activists tried to get on board but were thrown into the water by the *Senet* crew. Among the crew of the *Sirius* were two British citizens and a Dutch captain (N. N. 1994b; Bø 1994).

Later the same month, the Greenpeace ships *Sirius* and *Solo* and their rubber dinghies together with a Greenpeace helicopter followed the *Senet* in an attempt to block the vessel's hunt (N. N. 1994c). The Norwegian government passed a resolution to make it legal for the Coast Guard to intervene if whalers were disturbed by rubber boats, helicopters or by other means. Before this, the Coast Guard had been obliged to wait until activists physically 'attacked' whalers or whaling vessels before it intervened. Minister of Fisheries Jan Henry T. Olsen said the recent actions against the *Senet* had prompted the government's resolution, and *VG* called it 'the Greenpeace law' (Hansen 1994; see also N. N. 1994d). The *Sirius* and *Solo* disturbed the *Senet* again and their crews were arrested on 20 July and 23 July, respectively (Aass and Bergsaker 1994).

On 22 July 1994, fourteen Greenpeace activists from the Netherlands, Germany, England, Sweden and Italy were arrested and fined after erecting a wall with the message 'Closed due to whaling' in front of the entrance to the Ministry for Foreign Affairs in Oslo (Ramberg and Henriksen 1994). According to *Aftenposten Morgen*, charges were brought against several crew members from the *Sirius* and *Solo* who hailed from England, the Netherlands and the U.S., and they were fined 140,000 kroner in total (Bø 1994a, 1994b and 1994c; N. N. 1994e). In 1995, Greenpeace was ordered by a Norwegian court to pay the *Senet*'s skipper Arvid Enghaugen 300,000 Norwegian kroner to compensate him for the actions against his vessel in 1994 (N. N. 1995b; the 1994 Greenpeace actions against the *Senet* will be further discussed below).

In June 1995, sixteen Greenpeace activists from four different countries carried out a direct action in Tromsø against the whaling vessel *Havliner* which belonged to pro-whaling activist Steinar Bastesen. Some of the activists chained themselves to the vessel. Bastesen and the crew used a flame cutter and iron pincers to cut them loose. Some activists were thrown on the quay in Tromsø, but thirteen were still onboard the *Havliner* when the vessel left the port, and were later dropped off on Kvaløya (an island not far from Tromsø) (Johannessen and Bertinussen 1995; Sætra 1995; see also N. N. 1995a).

In May 1996, someone attempted to scuttle the whaling vessel *Elin Toril* in the Lofoten islands (Rapp 1996). Paul Watson claimed that Norwegians who sympathized with Sea Shepherd were responsible for the act (N. N. 1996b). Also in 1996, Greenpeace conducted a direct action against whaling in Ålo (Broch 2003), which will be discussed in more detail below. *VG* reported that a Greenpeace rubber boat that had been confiscated by the police after the direct actions against the *Senet* in 1994 would be given as a present to the Whaling Museum in Sandefjord (Kvistad et al. 1996). In April 1997, the *Senet* was reported to have been the target of an act of arson. A group which called itself 'Agenda 21' claimed to be responsible (Bjørnbakk 1997; N. N. 1997).

For the time between the 1996 Greenpeace action in Ålo and the implementation of a new strategy in the Greenpeace campaign against Norwegian whaling in 1999, I did not find any reports on further direct actions against whaling by Greenpeace in Norway. One of my Norwegian interviewees says that those who worked for Greenpeace at that time decided not to focus on the whaling campaign anymore:

> *Søren*: We just stopped doing anti-whaling campaigning, stopped sending out press releases, we stopped participating in the Greenpeace international actions, we stopped all that. We focused on fisheries and climate instead. ... We wanted the Norwegians to view Greenpeace as an environmental organization not an anti-whaling organization.

Thus, Greenpeace Norway was engaged in heated 'whale wars' for a number of years in the early to mid-1990s, then from 1996 all but abandoned the tactic of direct actions against whaling. What were the dynamics behind this?

The David against Goliath Chaos Communication Loop

The first chaos communication loop that emerged in the Greenpeace whale campaign was a loop in which both whalers and Greenpeacers acted on the basis of, and defended, their own collective self-descriptions. For each party, the actions and communications of the other party confirmed its own self-descriptions and reality constructions.

Nowhere was this more evident than during direct actions. In a paper titled 'The Battle of Ålo, 1996. A Descriptive Analysis of a Confrontation Between Greenpeace Activists and Norwegian Minke-whalers', Broch (2003) gives an account of how Norwegian whalers and members of whaling communities experienced a Greenpeace action. The whaling vessels involved were the *Feiebuen*, *Senet*, *Villduen* and *Feie*. They were anchored in Ålo harbour to unload whale meat at the storage point there (N. N. 1996c).

According to Broch (2003), the whalers at Ålo got wind of a planned Greenpeace action in advance. People from their area reported that they had spotted foreigners not too far from Ålo, and an inflatable rubber whale on the trailer of a car. The whalers also received calls from pro-whaling activist Steinar Bastesen and a journalist, informing them of a planned action (p. 180f.). Greenpeacers were aware that whalers often received such information through the grapevine, not least because individuals who worked for Greenpeace would be recognized if they travelled to whaling communities (see also Jonassen 1993). The whalers at Ålo were slightly nervous that the activists might damage their ships or their cargo, and contacted the Coast Guard and the police, as well as keeping watch on their vessels during the night (Broch 2003: 181ff.).

The direct action at Ålo took place on 14 June 1996. About forty activists from six different countries participated in the action (N. N. 1996c). Some activists tried to climb onboard the whaling vessels, and the whalers tried to stop them, spraying them with water hoses and using physical force. Communication between the activists and the whalers at Ålo mainly took the form of an exchange of standardized arguments. Activists said (among other things) that they wanted to put pressure on the Norwegian government to stop all whaling, that whales were endangered and that the whalers were violating international laws (Broch 2003: 185f.). Whalers stated (among other things) that their whale hunt was based on scientific and internationally accepted methods of whale counting, that it was sustainable and that it was not in their interest to extinguish the minke whale stock. There seemed to be little mutual understanding. 'Neither whalers nor demonstrators seemed willing to alter their expressed standpoints, not even an inch' (ibid., p. 186). Broch also mentions disagreements between different Greenpeacers involved (ibid., pp. 188f., 204).

Eventually, several activists managed to climb onboard, and the whalers became frightened that they might sabotage the engine or destroy the cargo (ibid., p. 187). This did not happen, however. The activists who were onboard chained themselves to the boats, while other activists chained themselves to the ropes between the boats and the quay. A few hours after, the activists unchained themselves and left; the police did not take anyone into custody (N. N. 1996c).

The whalers engaged in much storytelling and discussion of the events, much of it humorous (Broch 2003: 190f.). They spoke to their relatives on the phone and assured them that they were well (ibid., p. 194). While most of them agreed that they had 'won the battle of Ålo' (ibid., p. 192), some of them, particularly the elder whalers who were husbands and fathers, said they were sad about the incident. Broch quotes one of them as saying 'I am glad my father was not present. This was an attack on his very life-project' (ibid., p. 194; see also N. N. 1996c). They did not like the fact that they had become angry at the activists and had featured in dramatic media images, and feared that their families would think that they had behaved brutally (Broch 2003: 194), when they had always looked upon themselves as hard-working, peaceful family men (ibid., p. 202). They were touched and relieved that not only their families, but also many locals elsewhere in Norway whom they met on their next voyage, sympathized with them, even admired them.

In chats between the whalers and members of their communities, the Greenpeace activists were much ridiculed. There was even an amateur cabaret performance in one town depicting the direct action, in which the Greenpeacers were of course the butt of the jokes (ibid., pp. 194–197).

Broch's account illustrates the clash of opposed self-descriptions. Peaceful, idealistic protesters came up against peaceful, hard-working family men. Small Davids who were protecting nature came up against small Askeladens who were protecting their family's living and way of life, even their whole community's way of life. Greenpeacers wanted to stop all whaling because whales in general were endangered, and whalers argued that the minke whale hunt was sustainable and that they cared about sustainability.

Whalers were emotional about the confrontations with the Greenpeace activists and about their own self-descriptions:

> One whaler said right after the battle that if these demonstrators spoke the truth, he would have to rethink … everything he had learned and believed in during his whole lifetime. [Later,] he and the other whalers realized that such dramatic reorientation and renegotiations of personal and shared identities were unnecessary. … A positive self-esteem could be upheld by familiar ways of self-presentation. (Broch 2003: 204)

Greenpeacers and whalers saw each other as a frightening and potentially violent opponent whose arguments and point of view it was hardly worthwhile or even possible to try and understand (see also Kalland 2012, and the further accounts of Greenpeace actions below). This was in part because each party felt that their self-description was called into question by the confrontation, a frightening idea against which each party tried to psychologically defend themselves.

While whalers like those of Ålo basically defended themselves within their own local community against a 'surprise attack', pro-whaling activists took up the fight against the anti-whaling protesters in an altogether more professional manner and on the national and sometimes international level. Their self-descriptions, however, were the same as those that were in play at Ålo, given that they were whalers from whaling communities themselves.

There are several pro-whaling lobby organizations in Norway, the foremost of which is the High North Alliance (HNA). The HNA works to protect the right of whalers, sealers and fishermen to harvest ocean resources in accordance with the principle of sustainable management. It was founded in 1990 in the Lofoten islands as a result of cooperation by the Lofoten Regional Council (Lofotrådet) and the Nordland Minke Whalers' Union (Nordland Småkvalfangarlag) (Stenseth, Lid and Hoel 1993). Pro-whalers like the HNA portrayed themselves as a little Norwegian David who had lived with nature for centuries, understood her, and made its living from her. This David was threatened by a greedy 'protest industry' Goliath, which was trying to destroy traditional ways of life in Norway and make huge amounts of money for itself along the way (Helle and Stenerud 2003; Jonassen 1993). This protest industry was staffed by city people who knew nothing about the environment, based their arguments on emotions rather than facts, and had irrationally turned whales into symbols.

> 'Who are these urban environmentalists who are telling us how to run our lives? What do they know about us?' asks Steinar Bastesen, chairman of the Norwegian Small Whalers' Association. 'To us the whale is a fish like any other. We have families to feed'. (Lloyd-Roberts 1991)
>
> Georg Blichfeldt, a High North Alliance spokesman, ... said his tiny whaling community and others represented by the High North Alliance were dependent on whaling for their survival. ... 'These communities are dependent on whaling. I don't know if my society would have a future without whaling', Blichfeldt said. (Wallace 1992)
>
> The outside world has mixed up David and Goliath, the people here say.
>
> 'People in the U.S. ... think there's a big whaling industry, factory ships out there hunting down the last few whales. We're just a bunch of stupid little boats owned by families fighting for our rights. The only big organizations living off the whales today are Greenpeace and the Whale and Dolphin Society that use them as symbols for money raising [said Georg Blichfeldt, secretary of the High North Alliance]'. (Darnton 1993)
>
> 'Our problem is that in the past two decades the whale has become an urban totem', says Georg Blichfeldt, secretary of the High North Alliance. ... 'It is cultural imperialism to impose these foreign values on us'. (Maddox 1992)
>
> 'Look at this', [a representative of the High North Alliance] says. 'It's an advertisement prepared by Greenpeace Australia. The headline is "Are whales almost human? Parallels between a whale's life and that of our own make whale deaths even more disturbing", it says. "Whales live in family groups. An unquestionably intelligent creature, the whale has a gentle nature and is known to sing, play, even cry. Each whale has a distinct personality". ... They are creating a kind of super-whale with this emotive rhetoric. People think there's something called "the whale" when there are 80 different species. They are being led to think "the whale" is as clever as Einstein, as musically talented as Mozart and as holy as Mother Teresa'.
>
> 'But the minke is nothing to do with this myth'. (Barrett 1993)

The last quote captures the 'metaphysical' view of whales held by most Greenpeacers, which has been described in Chapter 1. John, a British former Greenpeacer, puts it as follows:

> *John:* Inside Greenpeace, there's the whale as a creature that's magnificent, deserves a special position in the environment, and we need to save this thing simply on the basis that it is ... magnificent. It'd be impossibly disgusting to lose it. So the blue whale is completely synonymous with the minke whale. Or any other whale. Because it's just – we must save the whale. And so the idea that there might be hundreds of thousands of one species of whale out there that is not threatened or is less threatened or that we don't even know how threatened it is ... doesn't fit.

Like the whalers, Greenpeace, campaigning against whaling in Norway, acted consistently according to its organizational self-descriptions of 'David against Goliath' and 'nonviolent protectors'. Seidl (2003) emphasizes that self-descriptions cannot be more or less 'true', 'right' or 'wrong'. An organization can produce multiple, even contradictory self-descriptions. However, self-descriptions can differ with regard to their degree of viability. A self-description can be judged viable as long as it fulfils its functions for the system. To be more precise: a self-description can be judged viable if it supports the continued existence of the system in its environment. This must include not treating the environment in such a way that the existence of the system itself becomes endangered (cf. Luhmann 1990). The Greenpeace self-description was certainly viable, for example, in the U.K., where it secured the environment's support for the whale campaign and for Greenpeace. But in Norway, the same self-description was nonviable.

In Norway, Greenpeace's organizational self-description meant that Greenpeace did not understand its opponents (whalers) or its audience (other Norwegians), their self-descriptions and reality constructions. Greenpeacers did not realize this, however, because the fact that Norway continued whaling confirmed its and its supporters' reality construction that Norwegians (at least, Norwegian whalers and politicians) were just stubborn bloody butchers who wanted to continue destroying nature for the sake of profit. Alternatively, they might be in the grip of a false ideology, and needed 'mindbombs' and a 'consciousness revolution' to be shaken out of it. Greenpeacers' self-description was thus also seemingly confirmed. Weyler (2004: 573f.) writes:

> 'When the Atomic Energy Commission or Japanese whalers called us crazy', said [Bob] Hunter, 'they were, essentially, right. We acted out like insane teenagers, exposing the family lie. It's not surprising the French reacted like an alcoholic father, beating up David [McTaggart, in 1973, J.R.]. In a dysfunctional family, the culprit always ridicules the whistle-blower'.
>
> We believed that in the face of society's psychotic behaviour – slavery, death camps, random murder, the fouling of rivers, decimation of whales, and so forth … – *someone* must stand outside the system, and perhaps break a rule to convey the unorthodox idea that might nudge society from its habitual course.

It can be safely assumed that most non-Norwegian (or at least non-Scandinavian) Greenpeacers took a similar view. Even some Norwegian Greenpeacers thought along these lines. As a consequence, Greenpeace continued to communicate in counterproductive ways.

> NORWEGIANS ARE LICKSPITS – … Campaign leader Pål Bugge of Greenpeace Norway [says that] most Norwegians are brainwashed, the authorities are incompetent, and Norwegian mass media are servile, that is bootlicking and lickspittling towards the power structure. … Bugge says that he can prove that the authorities lie with regard to the minke whale stock. … Asked for his opinion about the fact that most people in Norway support the resumption of whaling, he answers: You can't expect that average people will think something else than what they are fed with by mass media and authorities. (Schmidt 1992, my translation)

No matter how much a civil society organization may be in the right with regard to facts, it is hard to imagine it will gain much support from people whom its representatives call 'brainwashed lickspits'.

Greenpeace's own communication played into the hands of pro-whaling activists. Greenpeace campaigned to mobilize international outrage against and pressure on Norway. Boycotts against Norwegian products and organized letter- or postcard-writing have already been mentioned, and Greenpeace also published ads in leading British newspapers calling Norwegian whaling a 'massacre' and 'barbaric' (Andenæs 1987, my translation), as well as information materials using a similar tone. Such campaigning by Greenpeace and other groups delivered results abroad: for example, in 1993, the U.S. House of Representatives passed a resolution condemning Norwegian whaling, 347 to 0 (Darnton 1993). At the same time, this campaigning confirmed the pro-whaling activists' communication that these Goliaths, 'urban environmentalists' and 'cultural imperialists' wanted to tell Norwegians how to run their lives, wanted to keep Norwegians down and impose their own values on them. Indeed, the High North Alliance published a photograph of the 1994 Greenpeace actions against the Norwegian whaling vessel *Senet* on their website. The picture showed the Greenpeace ship *Solo*, chasing the much smaller *Senet*, attempting to spray it with water cannons. The HNA called the picture 'David and Goliath'. It was absolutely clear that the *Solo* was the Goliath and the *Senet* the David in this photo (see also Kalland 2012: 82f.).

Norwegian media quoted anti-whaling activists as stating 'If we can't save the whales, then how can we save ourselves' and playing tapes of whales 'talking' from loudspeakers (N. N. 1986f, my translation). They quoted Greenpeacers as insisting that the feelings people have for nature are important too (Straume 1992), and as saying that 'Food should be grown or reared. No one has the right to kill whales' (Spence 1996, my translation). Most Norwegians would see such statements as proof that Greenpeacers were indeed sentimental hippies and tree-huggers – or, to use John's term for it, 'Bambi lovers'. The pro-whaling campaigners could achieve powerful effects by simply presenting the anti-whaling campaigners' statements back to the Norwegian audience. The High North Alliance put an interactive game on their website which invited visitors to click on different arguments about whaling. If you clicked on the argument 'Whales are, because of their intelligence and gentle nature, unique and belong in category of their own that should not be exploited by man', for example, you would reach a picture of the 'M/S Born Free' and the 'M/S Intrinsic IV' lost in thick fog and the text:

> You are on board the Love Boat, the Cetacean Society International (CSI) vessel, lost in the mist of marine mammal myths. Your captain believes that whales 'are in a class by themselves' and that 'the two most highly developed forms of life (humans and whales) should coexist in peace'. (Robbins Barstow: 'Beyond whale species survival'). The captain of the Intrinsic IV, flagship of the World Wildlife Fund for Nature, also believes that 'whales have an intrinsic value as mammals of great intelligence, whose behaviour and language set them apart' (WWF about whaling…). Sorry, but the scientific literature does not support this view. While waiting for the mist to clear, read the essay: 'Brains, Behaviour and Intelligence in Cetaceans' by Margaret Klinowska or grab the life belt up in the corner.

The texts quoted were linked to the High North Alliance website, and of course, the arguments of WWF and Barstow would have been very similar to Greenpeace messages during the 1980s and 1990s. My interviewees also tell me that in response to anti-whalers' communication about the special intelligence of whales, people in Oslo would sport T-shirts saying 'Intelligent food for intelligent people' (see also Kalland 2012: 196).

As has already been mentioned, the Brundtland government allowed Norwegian commercial whaling to restart in 1993. In the statements Norwegian politicians made in support of Norwegian whaling, the 'David against Goliath' dynamic described here can be seen at work.

> 'Whaling is not actually economically important to us', says Helga Hernes of the foreign affairs ministry. … Instead, it is a question of the right to preserve a cultural heritage, the ministry says, arguing that 'Norway has always made its living from marine resources'. (Maddox 1992)
>
> 'It's important that small states … are able to resist pressure from major powers which, taking consideration of public opinion, disregard agreements and regulations for the management of natural resources'. (Johan Joergen Holst, foreign minister, quoted in Brown 1993)
>
> Prime Minister Brundtland said Norway had a long tradition of … using resources without depleting them. …
>
> 'Of course we feel dictated to', she said. 'It's a completely illogical, irrational, wrongly based campaign'. (Darnton 1993)

My interviewees are mostly convinced that Norwegian politicians' support for whaling was nothing but opportunistic. They think that whaling was never very important for Norway either in economic or in cultural terms. Their argument is that Norwegian politicians used whaling as a symbol of national autonomy etc. just as much as Greenpeace used it as a symbol of environmental destruction. In other words, Greenpeace and other anti-whaling groups gave Norwegian politicians the gift of an issue they could use to demonstrate attachment to Norwegian values and build political capital to spend on other issues, without incurring much political cost.

> *Sander*: There was the EU debate, whether Norway should apply for membership or not. And our prime minister at the time made the whaling discussion a kind of top-issue to demonstrate that, even if we would become an EU member, we would continue whaling. She made that kind of, the core definition of national self-regulation ability, even inside the European community.

> *Juliane*: So, she wanted to go into the EU, but in order not to be perceived as an anti-Norway person, she at the same time supported whaling.
> *Sander*: Yes. Because, as long as she was seen as a protector of whaling, the idea was that she would still be seen as a protector of Norwegian interests.

It is probably correct that the government wanted to win Northern Norwegian (or coastal Norwegian) votes (Krog and Talsnes 1992; Yngland 1992). But according to Grytås (2014: 335), who quotes an unpublished thesis by Hanne Kristoffersen and an interview with Brundtland's personal friend and former aide Lars Walløe, it seems there was also a psychological factor. Brundtland may simply have got angry because she was subjected to what she saw as undue international pressure in the whaling matter. It is thus quite possible that Brundtland and other Norwegian politicians had a genuine Norwegian 'David against Goliath' reaction against the anti-whaling protests and the international pressure.

The very things Greenpeace worked hard to achieve with their campaign were the things that were the undoing of the campaign in Norway. In the same way that Norwegians' actions confirmed Greenpeace's self-descriptions, Greenpeace's actions confirmed Norwegians' self-descriptions. Since for many Norwegians, international pressure to stop whaling was confirmation of cultural imperialism and oppression from outside, it was logical to respond to such pressure in the true Norwegian Askeladen spirit of resistance. From the Greenpeace perspective, the logical consequence of Norwegians' refusal to stop whaling was that Greenpeace had to continue campaigning against whaling, and if possible increase the international pressure on Norway. Greenpeace and their Norwegian antagonists got caught in a self-enforcing, self-reproducing cycle of mutual self-description confirmation and 'more of the same' actions (more pressure, more resistance, more pressure, more resistance…; see Watzlawick, Weakland and Fisch 1974) on both sides (see Figure 3.1). A leader in *Aftenposten Morgen* of 19 July 1994 discussing the 1994 Greenpeace actions against the *Senet* concluded:

> It's a source of continuing surprise that Greenpeace never tries to win the understanding of Norwegian public opinion for the view the organization represents. Many Norwegians have questioned the commercial minke whale hunt to start with. By provoking whalers who act in agreement with guidelines determined by the Norwegian authorities, Greenpeace does not achieve anything but to make the nation assemble behind the whalers who do their job. (N. N. 1994d, my translation)

The communication between Greenpeacers and Norwegians is an example of the systems-theoretic proposition that there is no such thing as a 'transfer' of information from one system to another in communication. Information is a system-internal quality. A system makes its own sense of things, not the sense that someone else might want it to make. Norwegians saw themselves as Davids under attack, no matter how much Greenpeace might have wanted them to see themselves as butchering Goliaths.

The case of Greenpeace in Norway is also a good example of how information is 'a difference which makes a difference'. There are certain differences which make a difference in a system. There are many other differences which do not make a difference in that system. And once some differences make a difference, other differences will not be information for the system. This is what Luhmann calls 'The start is fatal' (Luhmann 1997a: 111, my translation). Greenpeace saw itself as David, so the perturbations from the environment were interpreted in line with this. The idea that Greenpeace was Goliath (in the eyes of Norwegians, particularly Norwegian whalers) could not have been processed in the Greenpeace system and thus was not information in the system.

It must be noted that Norwegian Greenpeacers were aware of the counterproductive effects of the international Greenpeace organization's campaigning against Norwegian whaling. This understanding was one reason for the emergence of the other two 'chaos communication loops' that

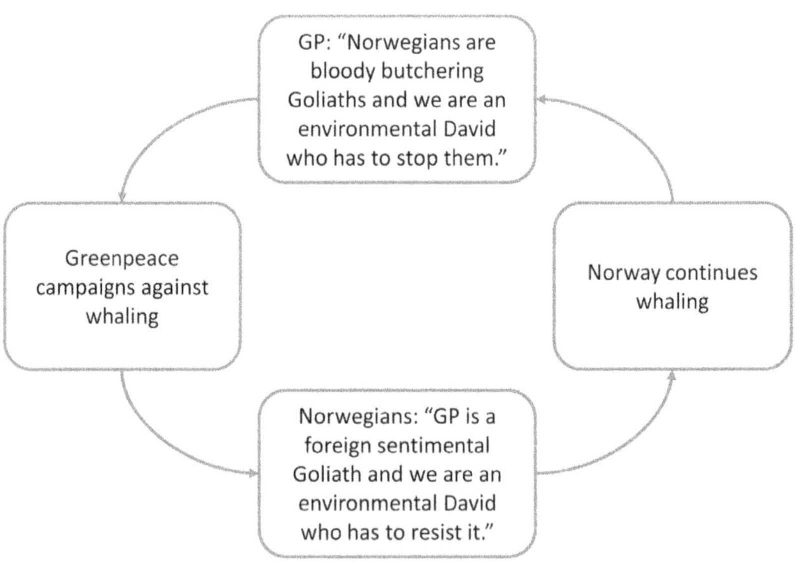

Figure 3.1. The David against Goliath chaos communication loop

characterized Greenpeace's campaign against Norwegian whaling in the 1990s. I call these the 'external and internal campaigning against each other chaos communication loops'.

The External Campaigning against Each Other Chaos Communication Loop

To understand how Greenpeacers were campaigning against each other in Norway, it must first be understood that the perspective on whaling in the Norwegian Greenpeace office was not as unified as it was in the rest of the international Greenpeace organization. Norwegian Greenpeacers held different views on whaling, although they all opposed it. Some were more or less pragmatic environmentalists and could, as Søren points out, also have worked in one of the other Norwegian environmental organizations such as Naturvernforbundet, Future in our Hands, or Bellona, which enjoy stronger support in the country. These campaigners opposed whaling, but they did not regard it as an important environmental issue compared for example to climate change.

Other Norwegian Greenpeacers, in particular those who worked on the whale campaign, were, to use Søren's expression, 'hardliners' on the whaling issue. Some of them, like Mikael, shared the international Greenpeace organization's view on whales and whaling. In his interview with me, Mikael says he just does not understand why this 'bloody, unnecessary slaughter' is kept up, and that to him, to stop whaling is 'almost like religion'.

In the eyes of many Norwegian Greenpeacers, the main argument against whaling was that there is great uncertainty as to how many whales out of a population you can kill without damaging the population. Whales are mammals, their reproduction is slow, and they are underwater most of the time and move long distances, so it is extremely hard to monitor how many whales there actually are. At the same time, as soon as a market and a whaling fleet exist, there will be an incentive to hunt threatened whale species pirate-wise and to lobby for increases in quotas and for permission to catch more whale species, because catch increases will more or less directly translate into profits. This makes whaling very prone to damaging whale populations. This argument is supported by economic models (see for example Clark 2005; Gordon 1954).

The quotas of minke whales Norwegian whalers were taking in the early to mid-1990s were probably not a real danger to the minke whale population, since the common minke whale is not viewed as threatened (www.iucnredlist.org). However, Norwegian Greenpeacers saw the argument as one of principle: as long as Norway allows whaling, this may encourage whaling in other countries, as well as pirate whaling, and may lead to overexploitation of both minke whale stocks and stocks of other whale species. (This is the reason why it is important that international agreements on the harvesting of endangered species are struck and their enforcement is controlled; see also Chapter 1.) Tor calls the Norwegian management with regard to whale catch quotas a 'historic scandal', and says it 'should never be trusted'.

Henrik, who was somewhere in the middle between the hardliners and the more pragmatic realpolitik-faction in the Norwegian Greenpeace office, puts more emphasis on the argument that Norwegian whaling was not in line with the International Whaling Commission agreements: 'And that can not be done, particularly not with a high-profile figure like Brundtland in charge, because that would undermine the international law systems'. The argument that Norwegian whaling was against international agreements (which will be further discussed later) was also put forward by Hans. Hans was not a hardliner about whaling at all, and was convinced that the Norwegian minke whale hunt was not dangerous for the whale population. But he found the Norwegian attitude to whaling 'narrow-minded':

> *Hans*: Minke whale hunting can probably be sustainable, but if the rest of the world is really, really concerned with this, we should take that into consideration. Norway should fulfil this commitment according to the moratorium. And if we don't do that, how can we argue that for instance the U.S. or other countries should follow the climate agreements and stuff like this, that hits their core industry?

Hans says he was frustrated because the whale campaign reduced Greenpeace's credibility in Norway and thus made it difficult to run other campaigns which he was more interested in: 'I certainly wanted the whale campaign as far away as I could get it, to be blunt. But, you know, it's an office, you have loyalty to the other campaigns in the office. And you run them. It is just a question of having to run things professionally'.

Henrik is very clear about that he was always convinced it wasn't really possible to completely stop Norwegian whaling, that the realistic possibility was to reduce and contain it. 'Norway knew so very well what fight they were up against when they made a decision to start. They knew this was going to create problems, and they were willing to take that fight'.

In short, although all Norwegian Greenpeace employees opposed Norwegian whaling, there was no such thing as a common perspective on Norwegian whaling within the Norwegian Greenpeace office. There were different personal motives and arguments regarding the whale campaign.

With some, there was a lack of belief that the whale campaign could actually be won, and unease about the scientific basis of the whale campaign and the negative impact it had on the credibility of other Greenpeace campaigns. The dominant perspective in the Greenpeace world (namely, the 'if we can't save the whales then what can we save?' perspective), in contrast, was much more uniform around the world than the Norwegian or Scandinavian Greenpeace perspective, which was and is rather a cluster of perspectives. (See Bailey 2009 on the different arguments against whaling.)

Being Norwegians, the Norwegian Greenpeacers understood why much of Greenpeace's communication about whales and whaling would be counterproductive in Norway. What some of them therefore tried to do, rather logically, was to communicate to the Norwegian audience those arguments against whaling that would better serve the anti-whaling cause. One-time campaigner Mari, for example, says that what she tried to advocate was for Norway to take a sustainable management perspective, with emphasis on the precautionary principle and on the fact that commercial whaling tends to result in over-exploitation of the stock. A holistic sustainable management perspective on this issue would have to take many variables into account: whales swim long distances, so the management would have to be internationally coordinated; it is difficult if not impossible to determine how large the stocks at a certain point in time really are (i.e., how many whales you may take out without damaging the stock); as whales are mammals, reproduction is slow; you have to take interaction with other species in the sea into account, which even today is an under-researched issue; other factors may harm whale populations in addition to whaling, etc.

This may have been a good argument to caution against whaling even from an average Norwegian's perspective. Most Norwegians oppose the unsustainable exploitation of nature (Grendstad et al. 2006: 107f.; Strømsnes, Selle and Grendstad 2009: 400). The first problem with the argument that Norwegian whaling would damage whale stocks, however, was that Norwegian scientists (who enjoy the trust of a majority of the population, see Grendstad et al. 2006: 133f.) contradicted it (Waagbø 1992). Norwegians trusted that their government would not allow whaling if it was unsustainable (in Chapter 4 I quote a study which confirmed this).

But an even bigger problem for Norwegian Greenpeacers was that while they were communicating about the unsustainability of Norwegian whaling, Greenpeacers elsewhere did not stop communicating the arguments that sounded mad to Norwegians. These mixed messages from the same organization robbed the attempts by the Norwegian campaigners to communicate in a way that was more sensible in Norway of all credibility they might otherwise have had. They made the Norwegian campaigners' arguments seem like a mere disguise for the underlying, Bambi-loving Greenpeace attitude, and a bad disguise at that.

> *Mari*: While we Norwegian Greenpeacers were arguing with the Norwegian politicians about sustainable management procedures, other Greenpeacers would argue from an animal welfare point of view, and then the next day it was all over the papers that 'Greenpeace says…', and it was only animal welfare argumentations. Then the politicians would say well, you're just into it because whales are cute, nananah, … and then you never got any further. *It was sort of like we [the Norwegian and the international Greenpeacers] were campaigning against each other.* (*Laughs*) Not that we did, but the world is so small, so if something is in the newspaper in the U.K., and it's correct from a U.K. point of view, when the same article is translated and put in a Norwegian newspaper, it is misunderstood. (Emphasis added)

That Greenpeace worked against each other internationally and destroyed much of each other's work is a judgment shared by several (ex-) Greenpeacers. It is evident in newspaper articles such as Omdahl (1993a). First, a Norwegian Greenpeace campaigner is quoted as saying that commercial whaling is not sustainable.

> It's this type of argumentation which drives [Greenpeace's] opponents crazy: While Greenpeace publishes 'obituary notices' for the whale and ads in which the whale is compared to a human being in a sentimental tone, [the campaigner] smugly claims that Greenpeace only argues on the basis of scientific facts. … Wouldn't it be just as well to admit that Greenpeace plays on the irrational and sentimental to win support? – Of course we play on emotions also. But first and foremost we use scientific, management and historical arguments [the campaigner says]. (My translation; see also Bentzrød 1992b; Mathismoen and Jonassen 1993; Borrevik 1993)

It was bad enough that the international Greenpeace organization kept up the communication that Greenpeace Norway was trying to replace with messages which would be more convincing in Norway. To make matters worse, the arguments Norwegian Greenpeacers were using were, in fact, at times inconsistent and incompatible with the arguments the rest of the organization was using. This of course damaged the credibility of Greenpeace in Norway – and of the whale campaign – further.

> *Hans*: I think most Norwegians will think that Greenpeace would have an argument with regard to whaling, if Greenpeace argued strictly on environmental grounds.
> *Juliane*: From your mouth that sounds simple enough. Could you argue on environmental grounds?
> *Hans*: No, because then we came into the loop where they said that: Well, fair enough, so you Greenpeacers are opposed to whaling because you say there are not enough whales for whaling to be sustainable, but what is the logical conclusion from this? The logical conclusion is, that Greenpeace would then have to accept whaling, if the numbers of whales in the population were high enough. And nobody in Greenpeace would want to do that. And that's the reason why Greenpeace has lost this campaign, because we have had mixed arguments all along. (See for example Sætra 1997 and High North Alliance 1997)

But even if Greenpeace's arguments had not been illogical, the Greenpeace whale campaign in Norway would still have been riddled with problems. This was because, ultimately, Greenpeace was sending out so many different messages to the media and the public that the impression it gave was rather chaotic. Greenpeace, as an organization, usually knows that only clear, simple, well-defined messages will win (cf. Murphy and Dee 1992). But the Norwegian Greenpeace campaigners did not have one single, worked-through, consistent argumentation, and so it was harder for them to speak with one voice than it was for campaigners for example in the Greenpeace U.K. office. And they got so frustrated with the continuous failure of the whale campaign, which undermined their other campaigns as well, that they grasped at every straw. They presented every argument against whaling they could think of: that there was a risk that it might overexploit and damage the whale stocks (N. N. 1991; Straume 1992; Myklebust 1993; Omdahl 1993a; Brøyn 1994); that it was against the IWC moratorium (Olsen 1991; Engesland et al. 1992; Hesstvedt 1992; Ramberg and Henriksen 1994); that the harpooned whales were subjected to an unacceptably long and painful death (Schmidt 1992; Mathismoen 1992b; Bø 1994a); that whales had become a symbol for the environment and people all over the world had strong emotions about the whaling issue, and that emotions were also important in the fight for the environment (Straume 1992; Bentzrød 1992b). Combined with the messages from the other Greenpeace offices, this made the Greenpeace communication about Norwegian whaling so polyphonic, in its entirety, as to make it very unlikely indeed that a majority of Norwegians would be convinced by it under any circumstances (see also Bragli Alstadheim 1992; Schmidt 1992; Waagbø 1992).

Figure 3.2. The external campaigning against each other chaos communication loop

Norwegian Greenpeacers developed strong emotions about the whale campaign. They felt frustrated and angry because of the continuous failure, anxious about confrontations with whalers, etc. These emotions were reactions to the fact that the Greenpeace communication was not viable in Norway, that the social system Greenpeace did not achieve a 'fit' with its Norwegian environment, had a disturbed relationship with its environment.

The Internal Campaigning against Each Other Chaos Communication Loop

If Norwegian Greenpeacers did understand the problems with the whale campaign in Norway better than their international colleagues, why did they never explain these problems to the rest of the organization? The answer is that they tried to at different points in time, but without great success. Their attempts at organization-internal problem-solving merely produced internal 'campaigning against each other' in addition to the 'campaigning against each other' with regard to external communication.

The Norwegian Greenpeacers did not have one single, worked-through argumentation in the whale campaign, and they did not have one single, worked-through approach with regard to tactics, either. Some of them mostly opposed putting international pressure on Norway through direct actions etc., as they saw it as counterproductive. Some of them thought that Greenpeace needed to keep up the pressure on Norway, because that was necessary to maintain international interest in the whaling question and to create opportunities to get the anti-whaling message out, even if it had a counterproductive effect too. One says that he saw that communicating different arguments against whaling in different countries caused difficulties in Norway, but thought that this had to be accepted. He argues that people are against whaling for different reasons in different countries, and that other Greenpeace offices needed to ensure support for the whale campaign in their country with the arguments that worked in their country. Some Norwegian Greenpeacers simply did not like the whale campaign very much and wanted to concentrate on other campaigns. One believed the whale campaign could not really be won. This combination of lack of motivation to invest energy in the whale campaign and disagreement about what would be an adequate strategy meant that the Norwegian Greenpeace office never spoke to international colleagues clearly and with one voice about the problems with the campaign.

However, organization-internal communication about the whale campaign also faced difficulties on a more fundamental level. Non-Scandinavian Greenpeacers, on the basis of their organizational self-descriptions, did not comprehend the problems Greenpeace had in Norway.

> *John*: The reason why the Norwegian Greenpeacers didn't explain to other Greenpeacers that they were campaigning against each other was, they didn't know how to tell them. They *could not* tell them. They would say the wrong things to them.
> *Juliane*: Like?
> *John*: They'd say, this is damaging our forest work. They'd say, you can't do actions because they upset people here. They'd say, the Norwegians think we're terrorists.
> *Juliane*: And then GPI would just go like, yeah well, deal with it.
> *John*: Yeah. Yeah. Absolutely. All of those things you have to deal with.

And at another point, John says:

> When I worked for Greenpeace in the U.K., I thought that the Norwegian Greenpeacers were just hiding. I thought they were being pathetic and stupid. I didn't give them the credit of having an argument. I thought it must be really tough. To be a Norwegian campaigner, it must be really hard. You're hated by everyone, but that's your job, and you should be sticking at it, and being strong about it. I never thought

> that it was impossible to do the job that they'd been asked to do because it was a dumb way to approach people. When they gave their excuses, even when they gave them in ways that I now understand, that they were trying to say: 'We should be doing it this way not that way', I took it as 'We don't want to do anything difficult'.

The reports from my Norwegian interviewees match what John says here: that Greenpeacers in offices outside Scandinavia often could not or did not relate properly to the Norwegian Greenpeacers' communications about their concerns about the whale campaign. Greenpeacers in offices outside Scandinavia were convinced that they were doing the right things to save the whales by undertaking direct actions against Norwegian whaling boats, by raising international pressure against Norway, etc. They had a well-defined mission and a clear strategic approach to the whale campaign. And in the framework of that approach, communication from Norwegian Greenpeacers which seemed to question the approach seemed, in Nils' words, 'mostly like distractions and excuses'. Again, we see here that communication is not a transfer of information. When Norwegian Greenpeacers told their international colleagues that Norwegians did not like direct actions against whaling, they meant that there was a fundamental problem with the whale campaign. But that was not the information international Greenpeacers understood.

We also see here what Niklas Luhmann meant when he said that both psychic and social systems are operationally closed and are separate from each other. Norwegian Greenpeacers had thoughts which did not agree with the communications of the social system Greenpeace. For example, they knew that Norwegians thought Greenpeace was Goliath. They tried to communicate this inside their organization. But such information could not be understood in the social system Greenpeace, and so was not understood. A social system is autonomous in its operations and cannot be determined by psychic systems, although it depends on them.

If the Norwegian Greenpeace office had not been financially dependent on the international Greenpeace organization, if the different offices had been more like equal partners, there might have been more of an incentive for Greenpeacers outside Scandinavia to try and piece the Norwegian Greenpeacers' complaints together. But as it was, Norwegian Greenpeacers' 'distractions and excuses' could easily be brushed aside by their international colleagues. As mentioned above, Greenpeace in the 1990s was no longer the Greenpeace of the 1970s which had been managed according to the 'Let a thousand Greenpeaces bloom' philosophy. It was a hierarchical organization in which power was relatively centralized and decision-makers mostly subscribed to the 'metaphysical' view of whales which was rooted in Greenpeace's countercultural history and tradition.

> *Alex*: Part of the reason why the problem did not get solved could also be a sense of inferiority in the Norwegian office because it was financially dependent on the international Greenpeace organization? That is something that small offices have to fight, sense of inferiority. The rest of the organization will not hesitate to push in your face, but then you should just be strong enough to say: 'We don't give a fuck. Close us down then'.

The non-Scandinavian Greenpeacers' certainty that they were doing the right things to save the whales was increased because most of the feedback they received from their most significant audiences – their supporters, media in their countries – encouraged them to stick to their strategic approach (see Kimura 2014 on Australian and Murata 2007 on British newspaper reporting on whaling; see also Kalland 1993). Many people outside Scandinavia had adopted the 'metaphysical view' on whales and expected Greenpeace to campaign according to it. As John, who worked in the U.K. office before moving to Sweden, says, 'When they see Greenpeace, people in countries like the U.K. think of whales. And there's a moral issue about accepting people's money and then not doing what they expect you to do with it'.

I have explained in the introduction that emotions may signal that the structure of the social system is viable in the environment, in other words, that the social system is doing fine and should continue like this. In other cases, emotions may signal that the social system is threatened in some way. Greenpeacers in the U.K., Germany or Australia received clear signals from their environments that their anti-whaling strategy, which used direct actions and 'metaphysical' (as well as 'rational') arguments and aimed to raise international outrage about whaling, was supported. They therefore had positive emotions about the way in which they were going about things.

Norwegian Greenpeacers, throughout the entire history of the Greenpeace whale campaign, appreciated the situation that their international colleagues were in, and the importance of the whale campaign for Greenpeace as an organization.

> *Hans*: There were people who thought that the whale campaign was important and that it was a Greenpeace campaign and a Greenpeace image. That we couldn't drop that, drop our history. And there is an element of truth in that. I can see that. It may be that it would be good for the Norwegian office if Greenpeace cut the anti-whaling campaign.

> But it wouldn't be very helpful, if the Dutch office, German office, American office, the U.K. office lost enormous support. It's just an example of why it is difficult to run international organizations...

On the other hand, the fact that non-Scandinavian Greenpeacers insisted on their approach to the whale campaign, and tried to get Norwegian Greenpeacers to support it, seemed like proof to some Norwegian Greenpeacers that their international colleagues were being somewhat opportunistic. All Norwegian Greenpeacers without exception state clearly that they believe that their international colleagues fundamentally believe that whaling should be stopped, that they are honest and authentic about it. But some of them say that Greenpeace's image and fundraising in countries such as Germany or the U.K. depended on the anti-whaling campaign. They say that this made the campaign, and specifically the 'traditional' strategic approach to it, more important and more long-lived than it would otherwise have been.

In Norway, of course, the anti-whaling campaign was not viable and not supported by the environment. Here, the emotions of the organization's employees – the Norwegian Greenpeacers – clearly signalled that the social system Greenpeace was threatened, that the system's structure of communication needed to change to be viable in the Norwegian environment. Like the strong positive emotions that international Greenpeacers had about the whale campaign, the negative emotions that Norwegian Greenpeacers had about it supported the social system Greenpeace.

The Norwegian Greenpeacers never successfully explained all the problems with the whale campaign to all their international colleagues in a comprehensive and (for the international colleagues) really understandable manner. And they did accept and support parts of the international Greenpeace approach to the campaign – some direct actions, some communication which sounded mad to Norwegian ears, etc. – because they thought they might be effective in some way after all.

But because they were sceptical about the campaign, and saw the problems, they also opposed some of their international colleagues' suggestions. I have explained in the introduction that organizational members' emotions may trigger contradiction, the communication of rejection, if the communication of a social system gets too far away from what its members perceive to be viable. Luhmann speaks of the immune function of contradiction, which brings with it the possibility of the system changing to new communicational meaning. Contradiction can help the social system to protect itself with the help of changes 'against rigidifying into repeated, but no longer environmentally adequate, patterns of behavior' (Luhmann 1995b: 371f.).

Direct actions, of the confronting, classic type the international organization wanted, were a particular point of contention. Norwegian Greenpeacers saw that such actions created a 'springboard' to get the anti-whaling message out, to be quoted in media reports, because the confrontation between Greenpeace and whalers was interesting to the media. But they also saw that these actions called forth the Norwegian nationalist Askeladen feelings described above, the feeling that Norwegians were environmental Davids against a sentimental Goliath. They therefore tried to prevent such actions more than once.

> *Mari*: We have had so many actions in Norway which totally failed and backfired against us. At the Lillehammer Olympics in 1994 they were thinking of having this huge whale fly over the opening session, they were thinking of filling it with something that looks like blood. Can you imagine the opening of the Olympics, you have all these elks and reindeers, and then you have this huge whale coming in and it's sort of exploding and you have all this blood on the elks and the reindeers and the snow and a sign that says: 'Norway stop whaling!'
>
> My argument was, the only thing it's gonna do is to ban Greenpeace from Norway completely. And you know, the Olympics, it's a sports-arrangement, it's supposed to be free of politics, a game where people are supposed to play and promote peace, and then GreenPEACE comes and ruins it because of politics? I said we're not doing it.
>
> *Søren*: The Norwegian Greenpeacers came to the conclusion, one, people wouldn't understand the connection between the Winter Olympics and whaling. Why? (*Juliane laughs*) And second, this was too sacred for the Norwegians, too sacred. Most of the people in the Norwegian Greenpeace office couldn't stand for this. They would have to quit. They couldn't defend it to their friends or anything. So it never happened.

An article about this action plan written by a former Greenpeacer appeared in *Nettavisen*, an internet newspaper in Norway, in 2006 (Stenerud 2006). Although the article was about plans for a Greenpeace action twelve years earlier which never actually took place, the piece was the most read article that day. This says something about the media attention that anything related to the Greenpeace whale campaign can get in Norway. Norwegians are still proud of the Lillehammer Olympic Games. A Greenpeace action at Lillehammer in 1994 would have confirmed all of the Norwegians' worst suspicions about Greenpeacers: that they were crazy and irrational (because what did the Winter Olympics have to do with whaling?), that they wanted to ruin everything Norwegians were proud of, etc.

In this case, the Norwegian Greenpeace office was successful in stopping the planned action. In other cases, it could not prevent the international organization from implementing its plans, and had to bear the resulting embarrassment. The following is an account of a direct action in Skrova in the Lofoten islands in 1993, provided by my Norwegian Greenpeace interviewee Mari (the text of the interview transcript has been summarized to make it more readable). At first, Mari tried, in Greenpeace-internal debates, to prevent this action. But she was overruled by Greenpeace England and Germany, and told to go to Skrova.

According to Mari, the Greenpeace activists (many of whom weren't Norwegian) were rather nervous just before the action:

> *Mari*: The whalers found out where in the whaling town we were staying like after half an hour after we had arrived there, and our Greenpeace people were so paranoid. I was like, if the police or anybody else want to find out where we are, they just need to go to the Greenpeace office and work as a volunteer, because Greenpeace is so open, everybody can go and work there, you know, so why are you so paranoid?

A report on the action in the national daily *VG* quotes a local saying that the action was no surprise for them because 'we had heard the day before that Greenpeace was in the area' (Nielsen and Stoltz Bertinussen 1993, my translation). Apparently, the activists' nervousness had been fuelled by the attitude of the Greenpeace managers who had planned the action.

> *Mari*: Greenpeace England and Germany had decided to have only female activists, because they thought the whalers could be violent. (*Both Mari and Juliane laugh*) I thought that was funny, I told them, you know, whalers are humans, they are like us, you should talk to them, not be so afraid of them.

The activists got even more frightened during the action:

> *Mari*: The day we had the action, we went out there early in the morning and all the girls chained themselves to the whaling vessels and I was chaining myself to one of the whaling vessels. And suddenly I got a call from one of the activists: (*speaks in a panicky voice*) 'You have to come over, the whaler is gonna kill us'. So I went over to them, and the whaler said: 'I was offering them coffee and cake' … (*laughs*) So I explained to the activists, these are regular people, they are nice people, they like whaling, but otherwise they are just like you and me, they're normal people, and they are not hostile.

> *Juliane*: But how could they, if he walks up to them with coffee and cake, how could they think….
> *Mari*: … I don't know. I think the situation, all the leaders were so paranoid, it sort of transmitted to the others, even if I tried to keep them calm.
> And it was so funny because this leading German campaigner, who was at the key site in Skrova, he actually fled! (*Laughs*) And flew back to Germany, he was so scared.
> *Juliane*: What did he think would happen to him?
> *Mari*: I don't know. The only thing that happened was some kids shooting with water pistols, that was the most dangerous thing. I think that it was because Northern Norwegians can be quite loud and yell at each other sometimes. And because of the paranoia of a few of the campaigners in the Norwegian Greenpeace office and also of other Greenpeacers, he was so scared that he chickened out.

The report on the action in *VG* is titled *Hjemmekoselig aksjon*, which translates as 'Homelike action' or 'Cozy action' and is unmistakably an ironic jibe at Greenpeace.

> For more than 14 hours they were doused with friendly offers of cream gateau, coffee and chipper whaling debate. Then the Greenpeace whale guerilla had had enough of Skrova and Friends of the Harpoon [an NGO started by housewives who show solidarity with whalers, see Kalland 2012: 104n8, J.R.]. … [The activists quietly] speeded out of the 300 souls whaling community in the Lofotens [in three rubber dinghies in the evening]. (Nielsen and Stoltz Bertinussen 1993, my translation)

The article contains some smug quotations from the whalers involved, and suggests that the Greenpeace action was relatively pointless. Mari's retrospective judgment is that she was right to try and prevent Greenpeace from undertaking this direct action in Skrova:

> *Mari*: You know, this situation where first, Greenpeace England and Greenpeace Germany wouldn't listen to reason and would go ahead with this action, and of course it backfired, 'cause nobody understood what we were doing up there and what we wanted to achieve. And then you had this campaigner who was fleeing… it was a little bit hilarious the whole thing.

Just days later, ex-Greenpeacer Bjørn Økern published his book, a scathing critique of Greenpeace (Bentzrød 1993; Omdahl 1993b).

Greenpeacers in Norway and Greenpeacers in other countries were disagreeing, discussing, debating, arguing back and forth, trying to stop each other from doing and saying some things or to make each other do and say other things. Søren tells me the story of how he was alone in the Greenpeace office in Oslo in 1997 and took a call from Associated Press. In his statement, he referred to the whale stock (*hvalbestand* in Norwegian). Shortly afterwards, his boss had an email from Lord Peter Melchett of Greenpeace U.K., demanding that Søren should be fired because he had said 'whale stock', not 'number of whales'. Using the words 'number of whales' would have made it clear that each whale is an individual. The story illustrates how pervasive the internal Greenpeace arguments had become. It seems Norwegian Greenpeacers were never able to move freely, nor were they ever able to get it right. They were under constant pressure from their international colleagues. Søren says himself that while the story is shocking in retrospect, it wasn't shocking to him at the time, because 'the climate in Greenpeace was like that… I even thought of myself, oh, maybe I shouldn't have said that'. But in retrospect, Søren sums up his judgment of the non-Scandinavian Greenpeace offices' approach to the campaign against Norwegian whaling as 'stupid'. This judgment is echoed by Sander.

As a result of the internal fighting, Greenpeacers outside Scandinavia thought that Norwegian Greenpeacers did not get the point, did not dare to

Figure 3.3. The internal campaigning against each other chaos communication loop

run direct actions because they did not want to be unpopular in their own country, were stubborn, etc. And Greenpeacers in Norway thought that Greenpeacers elsewhere were stubborn, sometimes opportunistic, did not get the point, did not want to listen to reason but then acted cowardly and chickened out of direct actions, and so forth.

In summary, we can say that the international Greenpeace organization pursued an action-oriented pressure strategy in the campaign against Norwegian whaling which used 'Bambi lover' arguments. Norwegian Greenpeacers differed amongst themselves on whaling. Some of them saw that the 'Bambi lover' strategy did not work in Norway, because it merely evoked (or at least helped their antagonists to evoke) feelings of Norwegians as being sensible environmental Davids fighting a sentimental Goliath who wanted to take away their culture and autonomy. Some Norwegian Greenpeacers therefore had negative or at least mixed feelings about the 'Bambi lover' strategy. However, Norwegian Greenpeacers were unable to communicate to the international Greenpeace organization that the 'Bambi lover' strategy was counterproductive. They were also put under pressure by the international organization to support that strategy. So they tried, in a piecemeal fashion, to object to certain plans internally, which led to constant organization-internal energy-consuming bickering. They also tried to communicate other arguments in Norway, which however made Greenpeace as an organization sound inconsistent, non-credible and cacophonic. Thus, different Greenpeacers destroyed much of each other's work. But it is important to understand that all the while they were meaning to do their best to make the whale campaign succeed.

The 'campaigning against each other loop' was self-reproducing and self-stabilizing because whatever action one of the involved parties took would

Figure 3.4. Internal and external campaigning against each other chaos communication loop

confirm the picture the other party had of it. The international Greenpeacers' lobbying for confrontational campaigning confirmed their Norwegian colleagues' view that the former did not understand Norway, were irrational about whales, etc. The resistance to confrontational campaigning from some Norwegian Greenpeacers confirmed international Greenpeacers' view that the former were hiding from difficult tasks, etc. And so on. Much of the time, there was no stimulus which would have led anyone to suspect that there was misinterpretation happening, that metacommunication might help, because the communication seemed to fit so well with the different parties' reality constructions.

To make the misery perfect, Greenpeace never achieved anything with its whale campaign in Norway. This allowed international Greenpeacers to point out to Norwegian Greenpeacers – and vice versa – that their wrong strategy/obstruction/hiding led to failure, and that if only they would see reason, the whale campaign would be OK.

Outcome of the 'Whale War'

The anti-whaling campaigning by Greenpeace and other groups was successful, over long periods of time, in the sense that many people outside Scandinavia came to see whales as special animals and whaling as inacceptable, and supported the anti-whaling cause. Yet there is no doubt that Greenpeace and other anti-whaling groups lost the 'whale war' in Norway. The most obvious evidence of this is the already mentioned fact that Norway defied the International Whaling Commission's moratorium on commercial whaling from 1986 to 1987 and from 1993 to the time of writing, and in 1988, 1989, 1990 and 1992, allowed whales to be caught for the 'purposes of scientific research' (Zangl 1999; Statistics Norway 2010). The fact that it was the government of Gro Harlem Brundtland, eponym of the Brundtland Commission and the Brundtland report, who defied the IWC moratorium, underlines how big a defeat this was for those who campaigned against Norwegian whaling.

Furthermore, as should be obvious by now, Greenpeace was quite unpopular in Norway. Greenpeace in Norway never managed to recruit a significant number of supporting members. Figures from Greenpeace statistics are 2,500 in 1990, 670 in 1994, 526 in 1997, 200 in 2000, and were up to slightly above 2,500 in 2005 again. This is blamed, by all interviewees, on the problems with the whale campaign. A 2002 survey carried out by the Norwegian Gallup Institute found that Greenpeace was very well known in Norway, but had less trust and support by far than any other environmental organization in Norway (Norsk Gallup Institutt AS 2002). It should be borne in mind that the 2002 Gallup findings came after major strategy changes in the Greenpeace whale campaign and after the 'whale wars' had become much less fierce. The outcome of a similar survey in the 1990s would probably have been even worse for Greenpeace. It is also reasonable to assume that some people associated Greenpeace with Sea Shepherd and other militant groups, getting the impression that Greenpeace is an organization which carries out violent actions. Greenpeace thus clearly failed in Norway along both lines of Gamson's (1975) definition of the success of protest groups: Greenpeace's beneficiary did not gain any new advantages, and Greenpeace never got accepted by its antagonists as a valid spokesman for a legitimate set of interests.

But Greenpeace's failure was even more comprehensive than that, in that it never managed to influence public opinion the way it wanted to. Most people in Norway were bystanders on the whaling issue with a leaning towards the pro-whaling cause, and there was very little support for the anti-whaling cause (on bystanders, see Gamson 2004). Grendstad et al. (2006: 109) report survey results according to which 'about 80% of the general Norwegian population said that to prohibit Norwegian … whale hunting was either "not very important" or "not important at all"'. They further quote a 1993 national public opinion poll according to which 'two in three of the Norwegian population said that the relaunch of the scientific [*sic*] hunt for minke whales – the Norwegian in-your-face position toward the international whaling commission as well as the international community – was a correct government decision'.

I have given some explanations of Norwegian national culture in Chapter 1. Many 'average' Norwegians would have found the whalers' communications convincing, since they appealed to and were in line with their sentiments and view of themselves and of Norway. Whaling was felt to be representative of Norwegians' right to autonomously secure their own sustenance, of Norwegians' heritage, culture and identity, of Norwegian sovereignty, and of Norwegians' freedom from political and cultural oppression. (See Blok 2008 for an analysis of Japanese pro-whaling countermobilization, which is similar to the Norwegian case.) It is not surprising that many Norwegian politicians ended up supporting whaling.

> *Hans*: It was never a real fight, because Greenpeace never really understood or even cared to understand how to win this fight in Norway. And I think that Greenpeace totally underestimated the opposition to interfering in Norwegian politics. Imagine if Greenpeace did the same for the climate campaign, as they have done for the whales.

Figure 3.5. All the chaos communication loops combined

> All this punch that this organization contains, and they used it to save 500 whales. Which they've lost. What a tragedy!

Luhmann (2006) predicts that it is improbable that organizations will exhibit a high degree of reflection in their day-to-day activities. This was obviously true for Greenpeace. Despite the abysmal results in Norway, the international Greenpeace organization just wanted to continue its business as usual in Norway in the 1980s and for the best part of the 1990s. According to Luhmann, organizations don't usually reflect very much because they are primarily busy with the continuation of their autopoiesis. If and when the system reflects, uncertainty is increased. Reflection makes life harder for organizations, because it produces awareness (even more awareness than there is anyway) that decisions taken might have been taken differently, that the organization could communicate differently if it decided to. Continuing operations becomes more difficult when every step is subject to questioning and discussion. In this sense, it is healthy that organizations do not reflect all the time.

Greenpeace exemplifies Luhmann's formidable problem description: if social systems can never 'see their environment as it really is', then what if they miss crucial points about their environments? For years, it was seemingly impossible for Greenpeace to change its communication structures and to thereby get out of the 'unproductive' whale campaign communication loops. Greenpeace went on reproducing itself without reflection, without realizing that it had missed important points, without realizing that its internal reality constructions were under-complex in a bad way. The international Greenpeacers of course received environmental feedback that was supportive of how they were communicating, and therefore felt emotionally attached to the way in which they were doing things. But the Norwegian Greenpeacers received quite different environmental feedback, and had quite different emotions – and these emotions did trigger contradiction sometimes – and still this was not enough to initiate fundamental organizational reflection.

Instead, the Norwegian Greenpeace office got caught in a pathogenic organizational double bind.

'Damned if You Do and Damned if You Don't'

The pathogenic double bind is a systems-theoretical concept originally developed by Gregory Bateson and colleagues.[2] A 'classic' pathogenic double bind situation consists of the following parts (the following builds on Bateson, Jackson, Haley, and Weakland, 1956 and Bateson 1983a):

- A relationship between two or more persons.
- One of the parties in the relationship communicates to the other a primary negative injunction in the form of 'If you do X (or if you don't do X) I will punish you'. This is often verbalized.
- This primary injunction is accompanied by a secondary injunction which conflicts with or denies the primary injunction on a more abstract level, and which also involves the threat of (often life-threatening) punishment (such as withdrawal of love and care). This is often nonverbal.
- Repeated experience: the structure of these conflicting injunctions becomes a habit, an expectation.
- An inability to leave the relationship: the parties feel that the relationship is very important to them in one way or another.
- Finally, metacommunication – for example, saying 'You are giving me two contradictory orders here; I cannot follow both' – is disabled.

Pathogenic double bind situations may be found in organizations (Kets de Vries and Miller 1984: 99ff.; Dopson and Neumann 1998: S63ff.; Tracy 2004; Cheek and Di Stefano Miller 1983; see also Argyris 1977: 116f.). A pathogenic double bind can even become a dominant feature of an organization, a characteristic of the system, turning all organizational members into victims (Hennestad 1990: 273; Soldow 1981: 509). Organizational members repeatedly receive mixed messages that they perceive to come not merely from certain individuals, but from their organization. To these discrepancies is added an understanding on the part of organizational members that by pointing out these discrepancies, they would break the rules, compromise themselves, etc. Organizational members can feel unable to leave the organization because they feel that their relationship with their organization is important to them. This can simply be because they believe they cannot find a job elsewhere, but more important in the context of this case are those organizational members who truly identify with the organization. Such individuals cannot easily leave the organization or at least distance themselves from it emotionally when the organization subjects them to contradictory messages. An organization like this is caught in a pathogenic organizational double bind (Hennestad 1990 puts it somewhat differently and speaks of 'double bind organizations'; see Figure 3.6).

The Norwegian Greenpeacers got caught in a pathogenic organizational double bind. They tried to follow two mutually exclusive injunctions, both of which they perceived to come from the same source – their organization. The first injunction was to gain the support of (average) Norwegians. As has

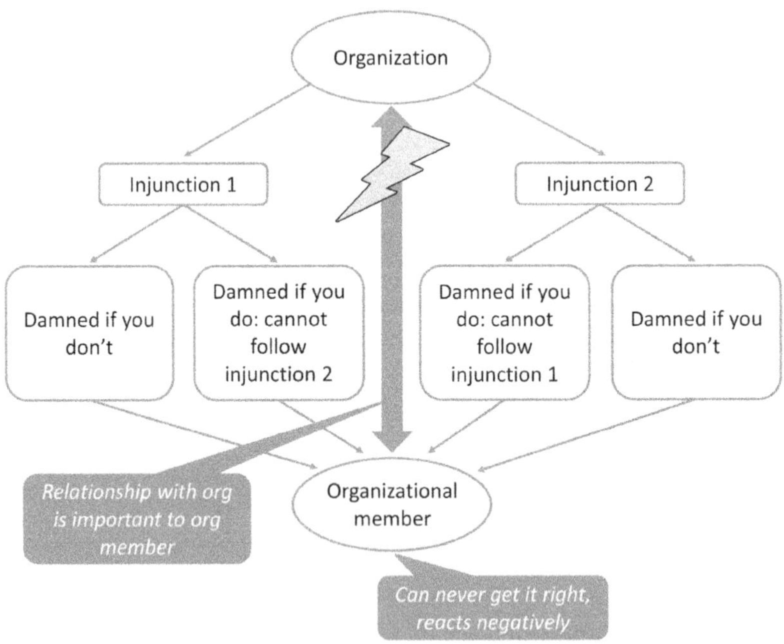

Figure 3.6. Pathogenic organizational double bind

been explained above, this goal was in line with the fundamental organizational beliefs of Greenpeace. Greenpeace only accepts donations from individuals, and sees its influence to be as great (or small) as the degree to which it and its goals are backed by society.

> *Hans*: The punch you have in political discussions, the punch you have when you speak with the people of influence, is based on the support of the people you have behind you. Indeed, I mean, Greenpeace agree with this. Which is why they are basing their support on individuals. So, why shouldn't that apply for Norway?

Norwegian Greenpeacers attempted to gain Norwegian support in different ways over the years, for example by campaigning on climate change and other issues which they thought would be important to Norwegians. Søren has been quoted above, saying: 'We wanted the Norwegians to view Greenpeace as an environmental organization not an anti-whaling organization'. However, the anti-whaling campaign always made it very difficult for Norwegian Greenpeacers to reach this goal. And even if they kept their mouths shut about whaling, Greenpeacers elsewhere obviously did not, thereby counteracting their Norwegian colleagues.

But Norwegian Greenpeacers also tried to follow their organization's injunction: 'Stop whaling!' As has been explained above, they tried to use arguments against whaling that were more convincing to Norwegians than the 'metaphysical' arguments that whales were special animals, etc. However, their argument that Norwegian whaling was unsustainable, for example, did not seem credible to Norwegians. And while Norwegian Greenpeacers attempted to argue in more 'Norwegian' ways, Greenpeacers in other countries did not stop communicating 'un-Norwegian' messages about whaling, thus undermining the Norwegian Greenpeacers' work. If Norwegian Greenpeacers then tried to communicate to their international colleagues that their approach to the whale campaign was problematic in Norway, the international colleagues perceived them as pathetic and as trying to avoid the task of running a difficult campaign. And then organization-internal bickering and quarrelling would ensue. Thus, the 'campaigning against each other chaos communication loops' emerged.

It was therefore impossible for Norwegian Greenpeacers to gain support for their organization, no matter how hard they tried, because Norwegians disliked them for wanting to stop whaling. At the same time, it was impossible for them to stop Norwegian whaling, because their organization did not enjoy any public support in Norway. The pathogenic organizational double bind that Norwegian Greenpeacers got caught in is illustrated in Figure 3.7.

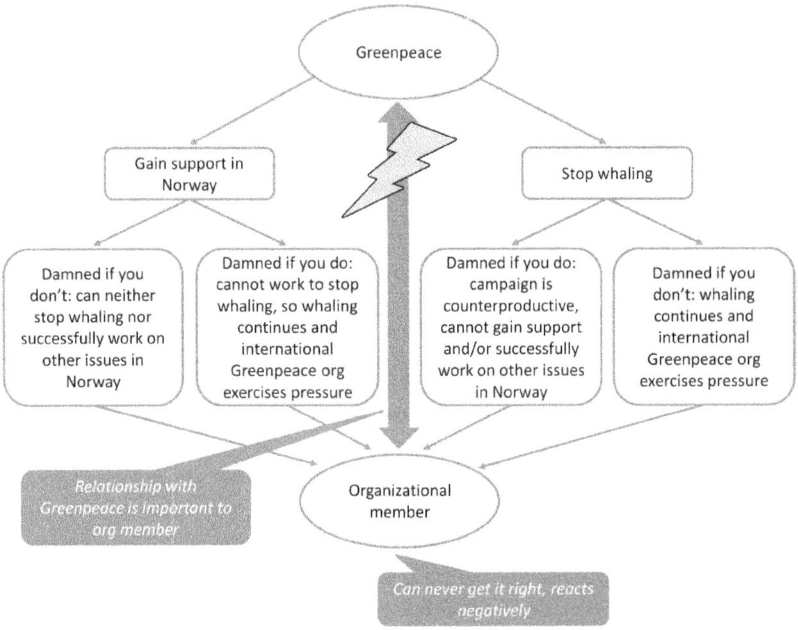

Figure 3.7. Greenpeace Norway's pathogenic organizational double bind

Over the years, the experience of 'being damned whatever you try' became a repeated one for Norwegian Greenpeacers. At the same time, they were very committed and emotionally attached to Greenpeace as an organization and to the organizational goals. Their relationship with Greenpeace was a crucial part of their social identity; they felt that their work was necessary; they could not simply pack it in and leave. It is possible that employees of civil society organizations working towards ideational goals are at greater risk of getting caught in pathogenic double binds than individuals working for other organizations, to the extent that they 'bind' themselves more to their organizations psychologically and emotionally and that they draw more personal meaning out of their work.

To complete the pathogenic organizational double bind, Norwegian Greenpeacers found metacommunication to be impossible. When they attempted to explain to their colleagues in other countries that 'people hate us for being against whaling' or 'people here think whaling is okay', their colleagues did not comprehend that they really were facing two tasks that were mutually exclusive. The non-Norwegian Greenpeacers would merely tell the Norwegian Greenpeacers to tough it out and repeat the message that whaling is not okay until Norwegians got it. They exerted pressure on

Norwegian Greenpeacers to do things in accordance with the classic campaign strategy, e.g. run direct actions. International Greenpeacers were unable to imagine that there could be valid arguments for whaling or against what Greenpeace was doing (at least for Greenpeacers). They might even communicate to Norwegian Greenpeacers that the latter compromised themselves by what they said.

> *John*: When I worked for Greenpeace in the U.K., I thought that the Norwegian Greenpeacers were just hiding. I thought they were being pathetic and stupid. I didn't give them the credit of having an argument. The Norwegians couldn't say the things that they really felt to the Greenpeace International team, because it was completely inappropriate for a Greenpeace person to say those things, as far as the International people were concerned. It took me a long time to understand what the hell was going on. Took me ages to realize that they had an argument. That it wasn't stupid, it wasn't utterly ridiculous, what the Norwegians were saying. Because I thought I was absolutely right.

Individuals who are caught in a pathogenic (organizational) double bind may feel confused (Bateson et al. 1956; Sluzki, Beavin, Tarnopolsky and Veron 1967; Leathers 1979), displeased (Leathers 1979), anxious (Bateson et al. 1956), frustrated (Dopson and Neumann 1998) or angry (Bateson et al. 1956; Sluzki, Beavin, Tarnopolsky and Veron 1967; Dopson and Neumann 1998) about the contradictory messages they are subjected to. Possible responses include: taking messages very literally and ignoring everything that is not literally said (Bateson et al. 1956; Dopson and Neumann 1998; Tracy 2004); constant over-analysis and looking for hidden cues behind what people said or did, i.e. paranoia (Bateson et al. 1956; Tracy 2004); ignoring the environment and trying to also be ignored by it (Bateson et al. 1956); oscillating between injunctions (Wagner 1978; Tracy 2004); defensive avoidance (Bateson et al. 1956; Janis and Mann 1979; Tracy 2004), and denial of personal responsibility (Tracy 2004; on the double bind theory see also Watzlawick, Helmick Beavin and Jackson 1968).

Since all or the majority of organizational members are caught in the pathogenic organizational double bind, the organization as a whole will be trapped between contradictory messages. It will switch from following one of the messages received, to following another, to trying to ignore the messages and so on. But it will never address the fact that there is a paradox, and will therefore not be able to search for the origin of the paradox or for constructive solutions to it.

Hennestad (1990) speaks of institutionalized learning incompetence (p. 278), where despite the fact that an organization seems to try something new or behave differently, this is not a real development. (Part of) what is really going on cannot be named and tackled; there is selective inattention, selective apperception, etc. to mute cues to the problem. Hence, the organization will be unable to adequately assess its own culture and the situation in which it finds itself with regard to its environment, and to adapt and (double-loop) learn appropriately (Argyris 1977; Hennestad 1990; Hirschhorn and Gilmore 1980; cf. Gabriel and Carr 2002: 357). It will be 'stuck' with its dysfunctional patterns of behaviour. As a consequence, the results of the organization's activities will be unsatisfactory. Organizational members will be frustrated about this and may begin to perceive all their experiences in double bind patterns, without being able to imagine that different ways of looking at what is happening might be possible. Their perception of their environment may therefore become fundamentally disturbed (cf. Bateson et al. 1956; Bateson 1983a). They may get caught in vicious circles of dysfunctional behaviour, which leads to unsatisfactory results, which leads to more dysfunctional behaviour.

Norwegian Greenpeacers exhibited the negative feelings that theory predicts people will have as a consequence of being caught in a pathogenic organizational double bind. They were anxious, not least because they received threatening anonymous phone calls in the dead of night, which they suspected came from supporters of the pro-whaling cause. They were frustrated and angry: '[Norwegians] call me Quisling, traitor and whore' (a Norwegian Greenpeace campaigner in 1992, quoted in Helle and Stenerud 2003: 10, my translation; see also Omdahl 1993a).

Norwegian Greenpeacers also exhibited the dysfunctional responses that theory predicts people will show as a consequence of being caught in a pathogenic organizational double bind. The first of these was something akin to paranoia. This requires some discussion.

It clearly transpires from both my interviews and my informal chats with Norwegian Greenpeacers that they often felt that Norwegian society stood together in favour of whaling, that Norwegians 'closed ranks' (Tor) against them. (Recall also the 1992 article in *VG* in which a Greenpeace campaigner was quoted as saying that most Norwegians were brainwashed and Norwegian mass media were bootlicking towards the power structure; Schmidt 1992.)

> *Mari:* When I called the Ministry of Fisheries, or the Ministry of Environment, it was really hard to get meetings, to get any information out, and if you got information out, it was blacked out, so you could

> hardly read anything, because everything that had to do with the whale or fisheries circuit was so sensitive.

Mari also says that the whalers 'had the Norwegian politicians and the Norwegian whale scientists to lean on'. Further, some Norwegian Greenpeacers describe the task of communicating with Norwegian journalists about the whale campaign as very challenging. There is even a hypothesis that there was clever government PR involved.

> *Henrik*: They lied from the top level with no problem. No problem to fool the whole population. The media is not controlled by the government in Norway. But when you hear things for so many years, you believe in them, regardless of whether you're a journalist or not.
>
> The government had a campaign. They decided to give out information in portions and in certain ways, to support their views. Always describing anti-whaling countries in certain ways, ensuring media had interviews with anti-whalers who stimulated that view. So, to have an interview with a British animal-welfare person was of course very positive for the Norwegian government. Instead of having a scientist in the U.K. who argued from a sustainable management point of view. You know they had their PR involved.

I do not know whether Norwegian journalists were prejudiced against Greenpeace, or whether the government influenced media reporting on the whaling issue in a pro-whaling, anti-Greenpeace way. However, the following points are probably worth bearing in mind when considering these questions.

First of all, media are prone to report more on an issue the more conflict there is around it, and to emphasize the conflict or event rather than to explain the underlying issue or condition (Gitlin 1980; see also Gans 1979). The whaling issue in Norway is no exception to that. All my informants agree that the conflict and confrontation between whalers and Greenpeace made the whale campaign a worthwhile thing to report on in Norway (and obviously, the reason why Greenpeace undertake direct actions in the first place is that they capture media attention, cf. Murphy and Dee 1992). It may thus be that Greenpeacers found some journalists hard to deal with because these journalists were more interested in the conflict than in the whaling issue as such (cf. Murphy and Dee 1992).

In addition to that, Norwegian journalists, obviously, were Norwegians and catered to a Norwegian audience. Some of them may therefore have been sceptical towards the Greenpeace whale campaign to begin with, for the reasons explained above. And some, particularly those working for regional or local media based in whaling communities, may have known that their audience was critical of Greenpeace. They may therefore have been inclined from the start to report on Greenpeace as a confrontation-oriented group which did not understand the whalers' perspective, etc.

On top of this, it must be borne in mind that Norwegian society is quite small. People with political influence – politicians, journalists, lobbyists – know each other better and meet each other more frequently than in bigger countries. As has been explained, Norwegian society is also quite homogeneous, with a traditional feeling that solidarity among Norwegians is a good thing. The hypothesis some Norwegian Greenpeacers voice – that social pressure played a role in the widespread Norwegian opposition against the anti-whaling protests – may therefore not be completely implausible (for a similar argument about Sweden, see Koch 1999). Apart from this, as sources such as Gamson, Croteau, Hoynes and Sasson (1992), Goodman (2004) and Gitlin (1980) tell us, it is obviously not only in small homogeneous countries that media content is homogenized, promotes quiescence, etc.

> *Henrik*: If you were a Norwegian journalist and you wrote negatively or critically about Norwegian whaling, you would be really attacked. From other media. Because it was like you were a traitor.
>
> This is a relatively small country, meaning there is a relatively small number of journalists and media. Even though we have diversity, it's not that many. Small number of government employees. Short connections. It's not a big network. Everybody knows everybody. If you go in on the whaling issue, it's so high-profiled. You don't dare to do that as a journalist. It's your career we're talking about.

The feeling of some Norwegian Greenpeacers, that Norwegian society 'closed ranks' against them and rejected them no matter what, may thus have been at least partly correct.

Yet, it is probably safe to say that the opposition to Greenpeace and to the anti-whaling protests among politicians, journalists and Norwegians generally was also a result of Greenpeace's chaos campaigning and inconsistent arguments. As mentioned above, even Leif Ryvarden resigned from his position as Chairman of the Board of Greenpeace Norway in protest against Greenpeace's direct action against Norwegian whaling in 1992, saying that 'it must be possible to harvest a renewable resource' (Waagbø 1992, my translation). Ryvarden was a natural scientist and was not opposed to Greenpeace per se. My interviewee Hans is completely blunt about his view that Greenpeace's arguments were not consistent or well enough thought through, and didn't hold water, and that this was why media reporting on Greenpeace's anti-whaling campaign was critical.

> *Hans*: The first thing I did, when I started working for Greenpeace, was that I took all the journalists who worked on environmental questions, out for lunch and had a few drinks and stuff like that, which I continued with for a long time. I had a very good relationship with many journalists, and I still do. The thing is, that there was a paranoia in this office about media. So many times I heard people say that the media are against us. But they're not, they are just not. I had very much good press on climate issues. The government was against us of course. I mean that's not surprising. But the idea that the government and the media cooperated is *ludicrous*. I think the Norwegian media expected to be able to ask some very critical questions of Greenpeace concerning the whale campaign, and I don't think we delivered. So they were more critical.

While Norwegian Greenpeacers' anxiety and frustration were understandable, they can also be interpreted as a paranoia-like response to the pathogenic organizational double bind they were caught in. Indeed, both Hans and Mari use the term paranoia at some point to describe some Norwegian Greenpeacers' attitude.

Another response to the pathogenic organizational double bind was that Greenpeace in Norway oscillated between the two mutually exclusive injunctions, 'Gain support' and 'Stop whaling'. Over the years, as has been described above, Norwegian Greenpeacers tried various tactics, arguments and approaches to do these two things in Norway, none of which worked. They campaigned on other issues than whaling to show that 'Greenpeace is something more than seals and whales' (N. N. 1990, my translation); then they protested against whaling again.

Yet another response exhibited by Greenpeace Norway was defensive avoidance of the whole problem. Søren has been quoted as saying: 'We just stopped doing anti-whaling campaigning in 1996. We focused on fisheries and climate instead. We wanted the Norwegians to view Greenpeace as an environmental organization not an anti-whaling organization'. It is very doubtful whether this strategy really would have led to Greenpeace gaining support in Norway, given that the international organization had not changed its stance on whaling.

Greenpeace Norway suffered from 'institutionalized learning incompetence'. Norwegian campaigners did try different things to solve their problems, but none of these constituted a solution. The organization was unable to assess adequately its own culture and situation and to adapt appropriately (cf. Hennestad 1990; Hirschhorn and Gilmore 1980). It was 'stuck', and this 'stuckness' persisted for years. As a consequence, as explained in the previous section, the results of Greenpeace Norway's campaigning were unsatisfactory. This only confirmed Norwegian Greenpeacers' negative perceptions, strengthened and entrenched the pathogenic organizational double bind, and gave rise to the self-reinforcing chaos communication loops described above.

Pathogenic Organizational Double Bind and Greenpeace's Self-descriptions

A pathogenic double bind can be interpreted as a defensive group mechanism or defensive relational pattern. It has been understood in this way particularly in the context of family therapy, where it has been extensively used (see for example Wilder 1979; Simon 2000b). Defensive mechanisms are processes and practices, such as projection or denial, that people employ in order to avoid pain, discomfort or anxiety. They are employed unconsciously and help to ensure that the threat or cause of the pain, discomfort or anxiety also remains unconscious; in other words, they defend those who use them against painful realizations (Freud 1967; Kets de Vries and Miller 1984: 134; Argyris 2006: 9). In so doing, they serve to protect the individual ego or the group (Freud 1967; Vince and Broussine 1996: 5). In and of themselves, defensive mechanisms are not pathological, and to some degree all individuals, groups and organizations use them to maintain their well-being and ability to function. However, exaggerated development or overuse of defensive mechanisms may have negative consequences (Kets de Vries and Miller 1984: 133ff.; Vince and Broussine 1996: 7; Argyris 2006: 9).

A pathogenic organizational double bind can then be understood as an organizational defensive mechanism or an organizational defensive relational pattern (cf. Vince and Broussine 1996: 4). It should not be understood as something imposed by one person on another, or by the organization on its members. Instead, all parties are bound and unconsciously (or at least not entirely consciously) bind themselves through the pathogenic organizational double bind (cf. Abeles 1976: 116; Bateson et al. 1963: 157). They do not metacommunicate or break the bind in some other way. They stay in the pathogenic organizational double bind because it covers up a conflict or problem in the relationship between the organization and its members that the organization and its members are (seemingly) unable to solve, and that would pose a threat to the organization if brought into the open (cf. Laing 1976; see also Papp 1980).

For organizational members, a pathogenic organizational double bind creates the illusion that if they succeed in sorting out the double bind, the entire problem will be sorted out. Therefore, they will spend their time attempting to deal with the contradictory injunctions of the pathogenic organizational double bind, rather than with the relational conflict that is covered up by it (see also Oakley 2000: 325). The relationship between the organization and the organizational members is thus maintained and stabilized. The organizational members are protected from the psychological discomfort that would be caused by confrontation with the conflict, and by the breakdown of their relationship with the organization to which such a confrontation might potentially lead (cf. Hirschhorn and Gilmore 1980; Wagner 1978; Willke 1999: 103f.).

From a psychodynamic perspective, this psychological discomfort can be intense because relationships in and with an organization can be as important to those involved in them as family relationships. Organizational members identify with the organization because they are searching for meaning, connectedness and empowerment, and are seeking to enhance their self-esteem (Ashforth and Mael 1989: 22; see also Diamond 1993). The identification with the organization may persist when the affiliation has personally painful consequences (Ashforth and Mael 1989: 34; see also Schwartz 2004 [1987]).

The pathogenic organizational double bind protects organizational members from the pain of realizing that their relationship with their organization is less than perfect, or of experiencing a breakdown of this relationship. At the same time, the organizational members will suffer because they are insecure about the validity or nature of the relationship (cf. Wynne 1976; Olson 1972; see also Wagner 1978).

Greenpeace Norway's pathogenic organizational double bind can be understood as an organizational defensive mechanism. It covered up the fact that there was confusion and contradiction in the relationship between the Greenpeace organization and Norwegian Greenpeacers with respect to what the organization wanted its Norwegian employees (and supporting members) to be and do. Norwegian Greenpeacers felt they should be People Persuaders, as Greenpeace is an organization that wants to change society and the collective consciousness and relies on the support of individuals for its campaigns. But being People Persuaders was difficult in Norway: while it might have been possible with regard to issues such as climate change, the anti-whaling campaign always got in the way of it. As a consequence, some Norwegian Greenpeacers felt that they should be Greenpeace's outpost, frontline, or extended army, in the High North, holding out against the whale butchers. This was a Norwegian version of the 'David against Goliath' organizational self-description: Greenpeace Norway was keeping up the fight against the Norwegian Goliath, with little support from 'average'

Norwegians. Over the years, some Norwegian Greenpeacers identified with the People Persuader organizational self-description, some with the Norwegian outpost or High North Frontline organizational self-description, and some tried to identify with both. Here, too, it seems Greenpeace Norway were unable to speak with one voice vis-à-vis Greenpeace International.

> *Henrik*: In my view, an organization like Greenpeace has a clear role to constantly drag society in the direction of being as environmentally friendly as it can. Greenpeace should challenge policy, challenge thinking, challenge behaviour in society. And that requires a certain confidence. And that's why I said, if more than ten per cent of the population like us, we didn't do a good job. If you start to be popular, and be mainstream, it's not your fucking job.

But Hans and Søren, when I asked them about this in my interviews with them, emphatically disagree with Henrik on this point.

> *Hans*: When I worked for the climate campaign, I felt that Greenpeace was too much on the fringe, I tried to draw it in to become more mainstream. Not to become mainstream, but to be more mainstream. Greenpeace is speaking on behalf of the environment. But who is to say what is best for the environment? It must be a group of people with strong public support. Indeed, I mean, Greenpeace agree with this. Which is why they are basing their support on individuals.
> *Søren*: Someone in Greenpeace said that our goal is to become unnecessary. And I like that. I thought the idea of Greenpeace was to caretake the earth, on behalf of future generations. And then you need to get people on your side. So I don't agree, I don't even understand the idea that if more than ten per cent of the population like us, then we are doing a bad job.

Norwegian Greenpeacers were emotionally attached to the different organizational self-descriptions. Let us recall that organizational self-descriptions enhance organizational members' self-esteem and offer them fulfilment and realization of self. They provide an organization with direction for its operations and a sense of its own unity and uniqueness. In other words, they fulfil very important functions (see Chapter 1). A Norwegian Greenpeacer who saw himself as part of Greenpeace's outpost or extended army would have taken the frustrations of the whale campaign not merely as part of his job, but as confirmation of his social identity, as something it was part of Greenpeacers' mission to endure. He may have derived a defiant sort of pride from it (see for example Schmidt 1992).

On the other hand, the self-description of being People Persuaders was also important to a number of Norwegian Greenpeacers. These individuals tried defensive avoidance of the whale campaign problem in order to work on other campaigns, or they tried to use more 'Norwegian' arguments against whaling. And some Norwegian Greenpeacers may have been torn over these different self-descriptions.

Norwegian Greenpeacers bound themselves to the pathogenic organizational double bind because they felt that the contradictions in their relationship with their organization were unresolvable. Their impression was that it was not possible to communicate meaningfully about the Norwegian difficulties with international Greenpeacers (see above on the impossibility of metacommunication). But they may also have been unwilling themselves to address the contradictions between the different organizational self-descriptions they were emotionally attached to. The relational problem between the Greenpeace organization and Norwegian Greenpeacers was fundamental; it was about the very question of what Greenpeace was. Solving it would have required a lot of time, effort and reflection, and potentially also a lot of conflict, and may simply have seemed to be an overwhelming task.

The binding to the pathogenic organizational double bind may have been quite unconscious most of the time, although some Norwegian Greenpeacers may have become conscious of parts of it at different points in time.

> *John*: There's no way a Norwegian Greenpeacer could have solved the problem. I'm absolutely certain about it. It's just too much. You end up fighting about the little stuff instead of the big one. 'Cause that's what human beings do. When you have a row with your boyfriend, you don't actually tackle his major issue, you go and talk about what he's done last week or the week before… And the closer they are to it, the more they go for the little bits. I think the Norwegian Greenpeacers can be completely excused for not seeing the big picture. And I'm sure on occasions they saw it. But those moments didn't happen to occur at the same time as a drive to make sure that it was communicated to people.

Norwegian Greenpeacers spent their time dealing with the contradictory injunctions 'Gain support in Norway' and 'Stop whaling', rather than confronting the problem in their relationship with the Greenpeace organization. This protected them from the discomfort of a serious conflict in or a breakdown of the relationship.

At the same time, Norwegian Greenpeacers suffered from the unhealthy relationship they had with their organization. They were 'stuck' with the inconsistent ways of relating and the dysfunctional patterns of behaviour, such as the internal campaigning against each other chaos communication loop. And in line with what double bind theory predicts (Bateson et al. 1956), Greenpeace Norway's unresolved internal relational problem led to Greenpeace Norway developing a more general problem of relating, also with its Norwegian environment. This became manifest in the David against Goliath and external campaigning against each other chaos communication loops. Of course, it increased the Norwegian Greenpeacers' suffering, as they constantly had negative feelings about the problematic relationship between Greenpeace – 'their' social system – and Greenpeace's Norwegian environment.

Systems theory tells us that reflection in autopoietic systems is rare and difficult. It is entirely possible that an autopoietic system will continue on a chosen path despite dysfunctional interaction with the environment, and never read the signs. The social system Greenpeace, however, was given a unique opportunity for reflection about its anti-whaling campaign when its regional chapter Greenpeace Nordic was established.

Notes

1. Parts of chapters 3–5 were published earlier in Riese, J. 2015. 'Taking Arms Against a Sea of Troubles: The Greenpeace Whale Campaign in Norway', *The Journal of Applied Behavioral Science* 51(1): 94–128. Republished with permission.
2. While it is beyond the scope of this book to give a comprehensive account of their research framework, it should be borne in mind (as Bateson, Jackson, Haley, and Weakland 1963: 155, insist) that the pathogenic double bind is not a stand-alone concept, but instead forms part of a more comprehensive approach.

Chapter 4

'Fuck Greenpeace, but Save the Whales'

Greenpeace Campaigning in Norway in 1998–1999

'Greenpeace Sweden Taking Control': the Merger into Greenpeace Nordic

The organization Greenpeace Nordic was the result of a merger process between the Greenpeace offices in Norway, Finland, Denmark and Sweden that took place between 1997 and 1999 (see Figure 4.1).

The merger of four offices into one, with subsidiary offices in Helsinki, Oslo and Copenhagen, and the main office in Stockholm, was a more or less direct consequence of a process in the international Greenpeace world during

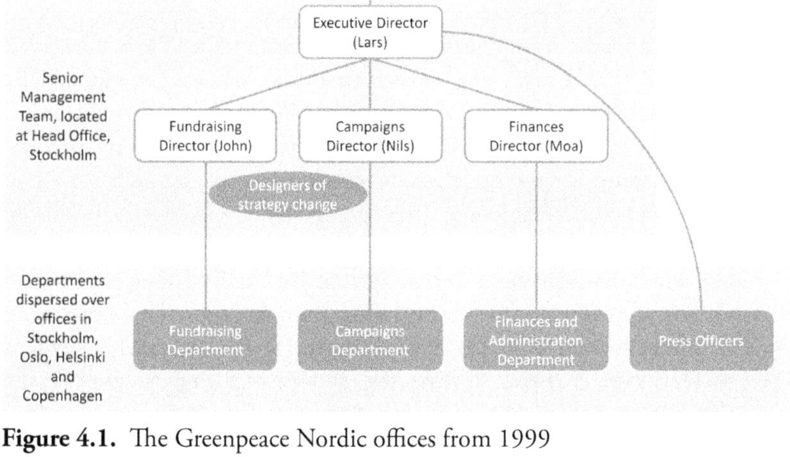

Figure 4.1. The Greenpeace Nordic offices from 1999

which national offices were requested to become self-supporting and contribute financially to the operations of Greenpeace International (GPI). Greenpeace International is an independent body based in Amsterdam. This process meant that a number of offices throughout the world were closed down because they could not meet the requirements. In some cases, offices were instead merged into regional offices. This process must be seen against the background of diminishing income for Greenpeace worldwide.

It transpires from interviews that Greenpeace International wanted to reduce its own administrative and cooperation costs by dealing with one regional office instead of several small ones in the Nordic region. GPI also argued that the Nordic region was comparatively small in terms of geography and population and that the work Greenpeace did there could be done by one, instead of four, organizations. The decision to merge Greenpeace Finland, Norway and Sweden into Greenpeace Nordic came about at a time when the Swedish office was (again) stable and dependable. It was in Greenpeace International's interest that Greenpeace Sweden took some responsibility for the weaker offices around it, and Greenpeace Sweden agreed with this. It is clear from interviews that Greenpeace International could (potentially) simply cut the financial resources that the offices in Norway and Finland depended on, which made for a relationship of unequal power.

Greenpeace Finland was angered by the merger decision. A conflict between national and international interests, similar to the one between Greenpeace Denmark and GPI over the fisheries campaign, erupted. In the end, Greenpeace Finland closed down the Finnish Greenpeace organization. This meant that Greenpeace Nordic had to go through a legal process to establish a new Greenpeace Finland, and get a new legal licence to recruit supporting members in Finland. Greenpeace Finland also lost all staff members who had worked there before the merger.

Unlike the Finns, most Norwegians, Swedes and later Danes accepted the argument that continuing several national offices was financially unattainable and that a merger was one possibility to meet this challenge. They saw the merger in a relatively positive light. But after the merger, the Norwegian Greenpeace office relatively quickly lost all the staff who had been employed before the merger. They saw the merger as effectively closing down the Norwegian office and being 'swallowed' by Greenpeace Sweden. It should be noted that Moa became the Greenpeace Nordic Finances Director, Nils became the Director of Campaigns, John became the Fundraising Director and Lars became the Executive Director. So the Senior Management Team – Heads of Finances, Campaigns and Fundraising, as well as the Executive Director – all came from the Swedish office. This was true also after the additional merger with Greenpeace Denmark. Norwegian

Greenpeacers felt they were not given enough autonomy in their campaign work; instead they were forced to travel to Stockholm all the time to have endless discussions about the campaigns. Out of frustration, they quit.

The merger process with Greenpeace Denmark in 1999 seems to have been more positive. Although it seems there was some resistance beforehand, people weren't too opposed to the merger when it actually happened. This is remarkable, given what has been reported above about tendencies towards 'national unilateralism' in the Danish office, as well as about the relationship between Greenpeace Denmark and Greenpeace International.

So in the beginning of 1999, Greenpeace Nordic had developed into an organization which could be characterized as follows. It was an international organization in structure. It was also an international organization in the sense that at least a number of people, including people in leading positions, placed strong emphasis on the idea of Greenpeace as an international organization. These people thought that it was necessary to balance an understanding of national cultures with international thinking and cooperation. Nils and John can certainly be counted among that number (just remember the Greenpeace Denmark story). Several people who had placed more emphasis on the importance of national campaigning had left. There was experience from different countries at hand in the organization: employees hailed from the Nordic countries, from Canada and the U.K. and had worked for Greenpeace in Russia and the U.S. Thus, there was a potential for increased organizational understanding of different cultures, including different attitudes towards the use of natural resources. Four-country Greenpeace Nordic had more weight vis-à-vis Greenpeace International and was, at least for the time being, relieved of pressing financial problems. Finally, Greenpeace Nordic was to a great extent shaped by two people – John and Nils – who were strongly convinced that Greenpeace should be People Persuaders. As Fundraising and Campaigns Directors, they brought the campaign philosophy that formed the basis of their work in Greenpeace Sweden to Greenpeace Nordic also.

John, the Fundraising Director of the new Greenpeace Nordic, had worked for Greenpeace in the U.K. from 1992 to 1995. He was an 'arch whale-lover' (his own words) and very much shared the 'If we can't save the whales, then what can we save?' perspective on whales and whaling that was predominant in the international Greenpeace organization. When he moved to Sweden and started working for Greenpeace there in 1995, he was in for some culture shocks.

> *John*: It was interesting to find that people didn't care about whaling. And actually had a very different attitude towards nature, in general.

> Someone had been given the Chairmanship of Greenpeace International, and the Swedes decided that they would give them a present. And the present they gave them was a stuffed moose head. And I mean, the inappropriateness of that as a present is so obvious to me, it's impossible. That was probably the first time I thought, there really is something going on here that's very different.

Gradually, through interaction with the Swedes, Finns, Danes and Norwegians working for Greenpeace, John realized that his view of nature, and particularly of whales, made him a Bambi lover in the eyes of many Scandinavians. When working for Greenpeace in the U.K., he could explain away Norwegian Greenpeacers' resistance to direct actions against whalers as being the result of their not wanting to be unpopular in Norway, etc. But now he was working alongside Swedish Greenpeacers who also viewed the whale campaign with some reservation and mixed feelings.

Björn tells me that at times, it was even as if the Swedish Greenpeace office ignored the whale question, closed one eye to it and hoped nothing would happen. Then, if a journalist would call, it was clear someone had to answer them, but nobody would really want to do it, because nobody wanted to be much involved and nobody had really good answers. There may even have been some reluctance to translate and send out press releases from GPI.

Greenpeace Sweden was also expected to help with direct actions in Norway, and these logistical support tasks weren't much more fun for Greenpeace Sweden than running the actions was for Greenpeace Norway. Nils recalls that 'it was perpetual nagging about this. It took a lot of energy from Greenpeace, these internal fights. We in Sweden had experienced this as long as I can recall in any case. It popped up every year, in a way, that we should do something, and then it was a lot of scuffle about what should be done and should not be done'.

In terms of national culture, the Swedes were somewhere in between the 'Bambi lovers' from the U.K. or Germany and the 'bloody butchers' in Norway (probably closer to the butchers than to the Bambi lovers). For this reason, it was easier for Norwegian Greenpeacers to communicate their concerns about the whale campaign to their Swedish colleagues than to colleagues outside the Nordic countries. John listened to the Swedish and Norwegian Greenpeacers' conversations and was exposed to their emotions about the whale campaign. It slowly dawned on him that he had missed crucial facts for a long time. He began to understand that the Norwegian Greenpeacers' concerns were legitimate and that Greenpeace as an organization had some real problems in Norway.

> *John*: The Norwegians within Greenpeace felt much freer talking to the Swedes about how they felt about whaling, than they felt talking to International. The Norwegians and the Swedes in Greenpeace could understand each other. It took me a long time to understand what the hell was going on. Took me ages to realize that they had an argument. That it wasn't stupid, it wasn't utterly ridiculous, what the Norwegians were saying.

Nils says clearly that his commitment to the whale campaign during his time as Campaign Director for Greenpeace Sweden and later Greenpeace Nordic was different from his commitment to, say, the forest campaign. Working on the whale campaign did not feel like addressing a pressing environmental problem, more like 'a necessary evil' or something that nobody else could do. Nils supported the whale campaign. But he would probably have chosen different issues to prioritize, if the whale campaign had not been such a problem for Greenpeace Nordic.

Yet Nils says he was the one pushing for the Nordic Greenpeace office to emphasize the whale campaign in 1999, because it was most important just then. He reports that he had seen for a long time before 1999 that Greenpeace had handled the whale campaign 'very badly', and needlessly antagonized people in Norway with it. At the same time, he saw that 'in so many places around the world, the whale campaign means a lot to people who work for Greenpeace or support Greenpeace'.

> *Nils*: I advocated for Greenpeace Nordic to focus on the whale campaign partly because I thought it was important to drive environmental work as Greenpeace in Norway and there was no chance of that if we did not first win the darn whale campaign.
>
> I also saw it as a chance to internationally, in Greenpeace, get the darn whale campaign sorted at some point. It had been and was a problem for Greenpeace internationally. People want different things and pull in different directions and debate and so a lot of energy is wasted without very much coming out of it.

Nils was so tired of wasting time and energy on internal debates and (at best) half-successful campaigning in Norway that he was ready to confront the problem.

John, of course, much more than Nils prioritized the whale campaign for the sake of the whales: 'Greenpeace without winning the whales, without saving the whales, would be a defeat. It wouldn't be Greenpeace proper'. Yet after working for Greenpeace in Sweden for several years, John, like Nils, saw that there was a real problem with the whale campaign and that Greenpeace would never be successful with any campaign in Norway until the whale campaign was sorted out.

Through working together as senior managers of Greenpeace Sweden and later Greenpeace Nordic, John and Nils had got to know each other quite well. They happened to like and respect each other personally. And so, Bambi Lover John and Butchers' Neighbour Nils sat down together and found that even if their motivations and perspectives were different, they could actually understand each other, and agree on some things. They were two people who could work as a team on the thankless task of improving the wretched Norwegian whale campaign. This 'allowed us to be focused in a way that we wouldn't have been if we'd been only one. I certainly wouldn't have done it alone. On my own, I'd never have done it' (John).

The Greenpeace-internal Argument: Reflection on the Greenpeace Self-description

> *Nils*: We did a strategy. And a lot of that plan we worked out in a kind of scuffle (*laughs*) between John and me. He came with something, which I discarded. And I came with something, which he discarded. And we didn't compromise, but we constantly tested ideas on each other. And every time one had tested an idea, which the other had discarded, still both changed their points of view, so that, in the end, it became a result.

John and Nils' discussions ran on over a considerable length of time. The main question they were debating was: what was the environmental issue at the heart of the whale campaign? John, who for a long time had subscribed to the 'metaphysical' view of whales, presented all the different arguments against whaling that he knew and argued that these were or should be at the heart of the whale campaign. Nils, who had never subscribed to the 'metaphysical' view, countered this, saying that an argument like 'whales are symbolic' could not and should not be the basis for the whale campaign. In this way, John and Nils worked their way through all the statements or ideas which could conceivably be regarded, by some people, as reasons to campaign against or to prohibit Norwegian whaling, and debated whether these were proper environmental arguments or not.

Nils and John ended up agreeing on the following. The real, environmental reason why the Norwegian minke whale hunt should end was that as long as one type of commercial whaling is legal, there will always be an incentive to encroach on other (possibly more endangered) whale species and to increase catch quotas, either illegally or legally.

> *John*: It's akin to the argument that you shouldn't have a market for ivory. You should not have a market for whale meat. In any way. Because it threatens other whale species that are threatened, there's a risk that other whales will be killed. Both immediately, pirate-wise, and in the long term, by encroachment into their markets and bigger market needs that would lead to expansion of the whaling industry.
>
> That's where we got to, Nils and I, and we believed that we could persuade others that this was the case, I could persuade Nils, and Nils could persuade me, and we'd get to the point where we could persuade the whole organization that this was the case. That, indeed, there was an environmental issue, and that it was important to persuade the Norwegians of it...

It is an important point that the two people who took on the job of developing a new strategy for the whale campaign of the new Greenpeace Nordic state that at the core of their debate was their reason for running the campaign. As will be seen, it was very important for the strategy which arguments were to be used in public, which tools were to be chosen, etc. These are obvious elements of any campaign which anyone developing a strategy will discuss. But Nils and John concentrated on why they would run this campaign, to start with. They were in complete agreement that they had to work that out first. As Nils puts it: 'I needed a motivation for myself to stand there and drive the rubber boat. I have to feel that it is right, you know, not just loose arguments, not just pretext. That I stand for this. Before we came that far we had quite a few decent arguments'.

The 'scuffle' between Nils and John was a reflection on Greenpeace's self-description. Let us recall that an operationally closed system (whether psychic or social) can never reflect on 'reality as it really is', because it can never see reality as it really is. It must use its own internal reconstructions of reality, including self-descriptions, as 'reality'. Nils and John wanted to find a viable self-description for Greenpeace in Norway. Who are we and why are we doing this? Why do we stand for this?

Nils' and John's 'scuffle' was also a constructive confrontation between two emotional worlds. John represented the world in which the 'classic' (countercultural and 'metaphysical') Greenpeace self-description and anti-whaling campaign strategy were buttressed by strong positive emotions. He represented the world in which emotions signalled the fact that Greenpeace as a social system was doing well and should continue its work. Nils represented the world in which emotions signalled the fact that there was a problem with the Greenpeace self-description and campaigning, that Greenpeace as a social system was not perfectly viable in its environment. John and Nils trusted each other and felt that both types of emotions sprung from true commitment to Greenpeace, that both types of emotions were valuable information, rather than proof that one of them was a pathetic idiot. They were exposed to each other's emotions over a period of time, and developed an emotional understanding of each other.

Nils' and John's 'scuffle' was cooked down to 'We are doing this because there is an environmental reason for being against the hunt of 500 minke whales a year, namely, the dangers of market dynamics'. But there was one aspect to Nils' and John's Greenpeace description which I think they never consciously discussed, because they knew they agreed on it and did not have to fight about it. This aspect is as important as the market dynamics argument. It is that Greenpeace should be a People Persuader. 'We believed that we could persuade others that this was the case, I could persuade Nils, and Nils could persuade me, and we'd get to the point where we could persuade the whole organization that this was the case. That, indeed, there was an environmental issue, and that it was important to persuade the Norwegians of it…'. John and Nils wanted to say and do things they themselves were convinced of, and they wanted to convince others of these things.

When the self-description of the Greenpeace system had been defined as People Persuaders with an environmental, rational reason, the way was free to look at how Greenpeace could persuade people that there are rational arguments against whaling. The self-description meant that Greenpeace needed a communication strategy explaining principles that are generally understandable to a majority of people. (It is a question of principles because it is not the killing of 500 minke whales that is the problem; it is the market dynamics that is the problem.)

The Argument for Norwegians: Understanding Norwegians' Self-descriptions

As the next step in 'sorting out the whale campaign', Nils and John set about finding out how Norwegians might be convinced that whaling should be stopped. They started this undertaking on the basis of the combination of their earlier professional experience as family therapist (Nils) and market researcher (John) and as campaigners on other campaigns (both of them), of their experiences with Norwegian Greenpeacers and Norwegians generally, and of common sense. They agreed that they needed to analyse the Norwegian situation and to produce a strategy that fit the Norwegian conditions.

> *Nils*: I was quite convinced, because I had dealt with systemic family therapy (*laughs*), and saw that Greenpeace was busy with, like, doing actions and trying to put international pressure on Norway, that that wasn't a method which you could win the campaign by. It had to be in another way. There is a mentality in Norway which makes that foreign pressure almost strengthens the resistance.
>
> When we did this campaign, we looked at Norwegian folk tales and so on. And discovered that all well-known Norwegian folk tales were about a little person, in a way, the youngest sister and the youngest brother, the poorest, the weakest. Who of course wins in the very end, over Evil and over his brothers and sisters, and never because he is strong, and always because he is smart. And there is a kind of funny nationalism, which isn't really that Norway is big and better than everything else. Norway was a poor country, almost a developing country before they found the oil. And always a kind of younger brother to Sweden, there is almost an inferiority complex in Norway. Which certainly clashes with when they realize that they are one of the richest countries in the world and have a lot of power, and that's a bit troublesome for Norwegians, I believe. It destroys their self-perception of being small and weak, but still fantastically capable.

> *John*: The next most important realization of obvious fact was that we were never going to beat Norwegians by putting outside pressure on. In fact, every time we tried to put economic pressure on them, it actually strengthened their resolution. The only way that we were going to persuade Norwegians to stop whaling was by having Norwegians decide that this is a bad thing to do. So we had to find the arguments that worked with the Norwegian public, not the ones that worked with the German public. And we couldn't carry on using arguments that worked with the German public, because it didn't matter whether the Germans felt that whaling's good or bad, because that wasn't stopping it.

In 1998, John ordered a focus group study from a professional researcher in Oslo on Norwegian attitudes to whale hunting, receptivity to arguments against whaling, attitudes to Greenpeace campaigning against whaling, and general knowledge of whales. Nils and John could not see 'reality as it really was'. But they could ask Norwegians what their reality construction was. They had begun to understand the views, emotions and self-descriptions of both Norwegian and non-Norwegian Greenpeacers, and found a self-description that they thought was viable for all of them. Now they wanted to understand the views, emotions and self-descriptions of Norwegian society.

The focus groups comprised men and women between eighteen and fifty-five years of age who were picked to represent 'average' Norwegian attitudes towards environmental protection and nature. John and Nils had made a list of all imaginable arguments against whaling. As part of the study, the researcher presented these to the participants and documented their reactions and the ensuing discussions. John was sitting behind a glass screen following the discussions.

> *John*: What we found was, they hadn't heard any of the arguments. Not one. All the communication we'd done, all the money that we'd spent, they hadn't heard any of the arguments. The good ones or the bad ones. They'd just heard a jumble of mixed message.
>
> They ended up with their opposition amounting to labelling Greenpeace as a terrorist group. A nonpatriotic, anti-Norwegian group. American, terrorist organization. Some really funny things, like, the boss of Greenpeace was a hairy lout who never washed and who had a prostitute for a wife. Absolutely strange. Urban mythology around it. And *when you started talking about Greenpeace, they ended up discussing what it means to be Norwegian. We were having a session about Greenpeace, and they were talking about deep-seated, important, meaningful stuff about what it was to be Norwegian.* And they were sitting there in a group discussion, reviewing the Askeladen, which is a Cinderella story, but with a boy, Askeladen being, Norway's being kept down by the Swedes and the Danes forever, and then suddenly, turn of the century, leapt free, and then what happened! (*talks in an awed voice*) Then they went to the South Pole, then they went to the North Pole, then they won all the big medals in skiing, and then they did this, and then they did… fantastic what a wonderful nation they are, we win the world, and, you know, we'll beat the British in football, and then we beat the whatever in something else, (*indicates the list going on*) tchchch and on and on and on, and this little nation that from afar feels like it's a solid European country that's been around for ever, turns out to be this little tiny country that's just finding itself, and, you know, then they found oil and oil means wealth and … it's like this little, hard-done-by country has made good and become great and fulfilled its destiny. And that I certainly didn't know beforehand.
>
> And none of the stuff that we'd ever done in terms of looking at Norwegians from a Greenpeace point of view had ever seen that. But what was happening here was, Greenpeace was an outside force trying to push them back to where they were before. Trying to say, you cannot

> do this, you do not understand. And they're saying, well we do understand. We understand there are enough minke whales to kill. And your arguments are completely bypassing that issue. There are our people, the Norwegians we believe in, saying that there are hundreds of thousands of minke whales. Your people are just saying, they're cute and cuddly and it hurts them when they die. This is natural! It is natural to kill a whale! (Emphasis added)

I will summarize the survey results (of which I obtained a copy) in what follows. The survey conveys quite a clear picture of the Norwegian 'environmental David against sentimental Goliath' problem. One result is that

> whaling is considered an obvious part of the Norwegian heritage among the participants in their 40s and 50s who grew up eating whale meat as part of their diet. … The participants in their 20s and 30s are more detached about the cultural significance of whaling, but have an abstract notion that it is part of the Norwegian heritage.

And, completely confirming everything Norwegian Greenpeacers said in their interviews with me:

> Many perceive that undefined foreign voices in the international community have a strong interest in the cessation of whaling for not-so-well-known reasons. … Norwegians perceive Norway as a small country that has been pressured by outside forces to end a regulated and safe hunt for whales. The majority feels that Norway should fight these forces because Norway has a right to administer its own resources and should not let itself be dictated on this matter.

According to the results summary, respondents are surprised to hear clear arguments against whaling, because '[t]he whaling debate has not left the focus group participants with one single convincing argument that would persuade them that whaling, as practiced by Norway today, should end'. But they find almost all of the arguments presented unconvincing, for reasons that are in line with what my Norwegian interviewees say, and with Norwegian political and cultural heritage. The only valid reason to end whaling, according to the focus group participants, is that the number of whales has become too low and that whales are threatened by extinction. Thus, when numbers have been restored, whaling can recommence. Some respondents 'express doubts that researchers are able to accurately determine the number of whales. Many feel that research (even Norwegian research) is often used to support diverse convictions on whaling and is therefore often untrustworthy. Nevertheless, many express confidence in Norwegian research agencies funded by the government'. This reflects Norwegians' trust in public institutions which had told the public all along that Norwegian whaling was sustainable. Respondents do not mind that whaling is subsidized, a result which reflects public approval of the government policies supporting a decentralized settlement structure.

As for respondents' perspective on whales, it became clear that Norwegians in general did not know much about whales and did not care too much about whaling per se. 'It is regarded as not credible, and if true, irrelevant information that whales are any more intelligent than other mammals'. (This is reminiscent of the Norwegian T-shirts with the slogan 'Intelligent food for intelligent people'.) The argument that minke whale hunting should be stopped because the minke whale is already under threat in a polluted ecosystem is discarded because there are many animals suffering from pollution and there should not be an exception for the minke whale. The argument that whaling should end because the whale is an animal with spiritual and mythological significance many people in other countries care deeply about 'immediately made people laugh'. As for respondents' view of Greenpeace, it became clear that they had a twisted perception of who or what Greenpeace was, mixed Greenpeace up with Sea Shepherd and simply disliked Greenpeace.

Respondents are equally clear in their views on how a successful campaign to stop Norwegian whaling could be set up. They do not believe Norwegian exports could be seriously damaged by Norway's whaling, and they believe that 'a campaign based on threats to the Norwegian economy would automatically put the audience into a defensive mode and close the discussion off right away'. While the respondents like the idea of whale safaris, they do not believe a campaign against whaling could be based on this idea, because 'there are no perceived contradictions between offering whale safaris while Norway is still hunting for whales', and it's 'not desirable to be invaded by tourists'.

That minke whale hunting in Norway encourages pirate whaling elsewhere is an argument the respondents 'take a considerable amount of time to understand'. When they understand it, they think it does not sound true because 'it must be difficult to catch, for instance a blue whale in secrecy' (it's too big), and because they do not believe there is a large market for whale meat. This is also the reason why the respondents do not think that the argument that Norwegian whaling increases pressure for a commercial whale meat market is a good one.

'The respondents agree that the best argument to use in a campaign, that seeks to end whale hunting, is to inform Norwegians that whale hunting is not permitted by international agreements and laws'. Few of the respondents are aware that Norway is breaking international treaties by hunting whales. The obedience to international law is judged more important than their personal convictions. 'If Norway is breaking an international agreement, many feel that other countries will respect other international agreements less as well. Several suggest that Norway should cease the whale hunt and as a second step work to alter the international laws that forbid whale hunting'.

With regard to the Greenpeace argument, presented to the focus group respondents, that Norwegian whaling violates international agreements, it should be noted that Norway argues that its commercial whale hunt is legal. This is because Norway made use of its right to file an objection against the IWC moratorium. Norway further argues that even those in the IWC who are opposed to whaling accept that the Norwegian hunt is legal (Johansen 1999). Greenpeace argues that the fact that Norway's objection is permitted in the 1946 International Convention for the Regulation of Whaling is not relevant in this case. According to Greenpeace, a distinction must be made between strictly following the original Whaling Convention from 1946 and actually cooperating with the IWC with a view to the conservation of marine mammals – which is what is required in the United Nations Convention on the Law of the Sea (UNCLOS).[1] I will come back to this point below; for now, suffice to say that Greenpeace took the view that Norwegian whaling was against international agreements (see also Lysvold 2014).

The bottom line of the focus group study was thus as follows. The focus group respondents would be concerned if minke whales were brought in danger of extinction. But they did not believe this was the case in Norway, because they believed the Norwegian researchers, who had been saying for years that the Norwegian minke whale hunt was safe. This of course confirms what Norwegian Greenpeacers said in interviews about their fellow countrymen: that they would mind if whaling endangered whales, but that they would not accept it if Greenpeace said 'and even if whales were not endangered, whaling should be stopped anyway'. The only argument against Norwegian whaling the focus group respondents accepted was that Norway should not violate international agreements.

Concern about respect for international agreements is perfectly consistent with the Norwegian 'David against Goliath' attitude. Norway is a tiny country with little military or other power, which has suffered in the past because bigger countries did not stick to the rules. Such a country will want everybody to stick to the rules and will be anxious to do so itself. Finally, the focus group study participants clearly opined that a campaign against whaling 'should present scientific facts about whales and whaling *from an independent and trusted source*, and should avoid arguments based only on emotion' (emphasis added).

The focus group study made it possible to derive clear instructions for the whale campaign. If the Norwegian environment thinks international agreements should be respected, then Greenpeace should highlight what it regards as Norway's breach of international agreements about whaling. On the other hand, if Norwegians have a very negative picture of Greenpeace, if they think Greenpeace is strange and irrational, then Greenpeace is not the rational voice that can tell them to stop whaling.

With their discussion about 'their' reason for campaigning against whaling and with their focus group study, Nils and John were striving for reflexivity, in the systems-theoretic sense. Instead of just taking the existing Greenpeace-internal reality constructions at face value and assuming that this was reality, Nils and John tried to find out whether the constructions could be corrected, improved. To do this, Nils and John needed to find out how the environment 'ticked'. They needed to take seriously the 'warning' emotions Norwegian Greenpeacers had had about the anti-whaling campaign all along and to develop an emotional understanding of Norwegians.

Through the focus group study and reading the Askeladen tales, John and Nils came to understand the 'environmental David against sentimental Goliath' problem and gathered clear hints as to how to redesign the whale campaign. They were now ready for a critical analysis of Greenpeace's anti-whaling campaigning of the past.

Analysing Former Campaigning: Understanding the Loops and Binds

Nils and John tried to identify the problems with the strategy which had been in use before. They were aware by now that Scandinavian and non-Scandinavian Greenpeacers lived in different emotional worlds, and how much their perspectives on whales and whaling differed. Nils and John acknowledged that the international Greenpeace organization had never convinced the whole Norwegian Greenpeace office (nor, for that matter, the whole Swedish or Danish Greenpeace office) to support the whale campaign as staunchly as the rest of Greenpeace supported it. International Greenpeacers simply never grasped that their own perspective on whales was not as obvious and convincing for all Norwegian Greenpeacers as it was for them. As John puts it: 'They were coming from the Bambi lover position I had before'.

John and Nils had understood from the focus group study that the Norwegian public had never heard any distinct arguments against whaling, but only 'a jumble of mixed message' (John). They could now interpret why this was so.

> *John*: Before '98, the Norwegian Greenpeace office basically confused Norwegians by putting out messages that were completely mixed. They had messages that they were forced to do from International, they had messages that they themselves made up about rational reasons why you shouldn't do whaling, they had a whole series of different messages. They ran the logistics for actions they didn't believe in. They stood in front of cameras saying things they didn't agree with. It was a mess. The communication of the Norwegians was a mess.

Nils and John tried to get a comprehensive picture of the political situation in Norway, on which of course the focus group study had already shed some light. They quickly understood that many politicians in Norway supported whaling because it made them popular in coastal communities and possibly also among other groups and basically could not cost them any votes. Nils and John looked at the communication of the pro-whaling campaigners, played the boat game on the website of the High North Alliance and so on. They saw the picture that was displayed on the HNA website under the heading 'David and Goliath', of the big Greenpeace ship *Solo* trying to spray the small Norwegian whaling vessel *Senet* with water cannons. They saw how the pro-whalers simply presented back the divergent arguments against whaling put forward around the world in ways that made these arguments look ridiculous. John and Nils realized that their Norwegian antagonists were in fact campaigning quite skilfully, using good arguments that resonated with their audience, and that Greenpeace had not been counteracting this at all well. Putting together the information they had, Nils, who had worked as a systemic family therapist earlier, was able to recognize the 'more of the same' pattern at work in the David against Goliath chaos communication loop. Greenpeace mobilized international pressure on Norway to stop whaling. Although Norwegians did not care much about whaling in itself, this challenged them to resist. Then Greenpeace saw that its campaign did not work, so it believed the pressure had to be increased. This challenged Norwegians to resist even more. And so on.

Nils also understood that as a result of the failure of the campaign, a similar pattern emerged inside Greenpeace and consolidated into the internal campaigning against each other chaos communication loop. Norwegian Greenpeacers saw that the campaign did not work, so they tried to prevent international Greenpeacers from doing things they thought would be counterproductive in Norway. This made international Greenpeacers increase the pressure on Greenpeace Norway to campaign in Norway according to the 'classic' strategy. This made Norwegian Greenpeacers more fearful of further failure and more desperate to prevent that type of campaigning. And so on.

> *Nils*: What Greenpeace Norway had done before was like, to keep the other Greenpeacers away all the time. Because those who worked as campaigners in Norway saw what happened. 'More of the same', only positive feedback loops all the time, whatever Greenpeace did, things got worse and worse. But they wanted to solve the problem through getting rid of it, you know. If only those international Greenpeacers stop coming here and spoiling things all the time, then everything will be good.
>
> And that didn't really work either, because it also becomes a kind of 'more of the same'. Then the international Greenpeacers think: Damn idiot Greenpeace Norway who don't understand anything.

Nils and John realized that Greenpeace in Norway had got caught in internal and external chaos communication loops and trapped in a pathogenic organizational double bind: 'campaigners in Norway saw that whatever Greenpeace did, things got worse and worse' (Nils). And they also realized that it had been impossible in the past to break out of these loops through metacommunication. John has already been quoted on this.

Nils says that one simple partial explanation for the lack of metacommunication was that Greenpeacers from the Norwegian office and those from offices in the U.K., the Netherlands, etc. did not meet very often. Of course they met sometimes and communicated with each other, but they did not get together long enough and often enough for the non-Scandinavian Greenpeacers to become aware of how big the problem really was. The non-Scandinavian Greenpeacers never went through the process John had gone through after moving to Sweden and seeing Greenpeace Sweden give the Chairperson of GPI a stuffed moose head for a present. (Even the Chairperson of GPI who got the stuffed moose head would probably have brushed it aside as an incidental Swedish eccentricity.) Non-Scandinavian and Norwegian Greenpeacers were not sufficiently exposed to each other's emotions. Because there was a lack of emotional understanding between them, there could not be sufficient understanding of their respective arguments between them.

Juliane: When the Norwegians explained to GPI that they wouldn't achieve anything in Norway except that Greenpeace becomes unpopular, didn't GPI grasp that?
Nils: But then we're back on to that question we were on before that okay, maybe we don't achieve so much in Norway. But maybe we achieve something for the whales. Then the Norwegian Greenpeacers say No. Then GPI say Yes. And then you keep discussing. And you never really got to know, because both sides spoilt each other's work in a way. (*chuckles*) Worked against each other.

What is remarkable about John's and Nils' analysis of earlier campaigning is the respect and the understanding they express for the Greenpeacers involved. It would be easy for them to accuse their colleagues of stupidity or lack of effort and to just generally blame them personally for the failure of the whale campaign. They could also attribute that failure to the Norwegian public being deliberately obtuse, or take some other easy way out. They do none of this. John has been quoted at length as saying that he thinks that Norwegian and international Greenpeacers simply could not understand each other, even if they all spoke English. He has also been quoted as saying that he thinks that Norwegian Greenpeacers were 'too close' to the problem to be able to see 'the big picture', and that on the occasions they did see it, they were unable to communicate it to the rest of the organization. Nils takes a similar view.

And indeed John and Nils are right in not blaming their Norwegian colleagues. There were Norwegian campaigners who understood that it was a problem that Greenpeace used inconsistent and unclear arguments in the campaign against Norwegian whaling. There were Norwegian campaigners who understood that while Greenpeacers saw themselves as environmentalists and Norwegians as butchers, Norwegians thought that they were the real environmentalists and Greenpeace was a bunch of Bambi-lovers. There were Norwegian campaigners who understood that it was a problem that Greenpeace did not agree internally on why whaling should be opposed. There were those who knew that Norwegians would be against whaling if it could be established to be a real danger to the whale populations. All Norwegian Greenpeacers were aware of the risks and the importance of direct actions.

Most striking of all, Norwegian Greenpeacers knew that Norwegians would buy the argument that Norway should not violate international agreements. They were convinced that this argument was a good one, years before Nils and John ordered the focus group study. Henrik says that Norwegians do not believe that the Norwegian minke whale hunt is unsustainable, because they have such trust in the Norwegian government, scientists and other institutions: 'But the international conventions they buy. It doesn't criticize anybody, it doesn't take away the belief in Santa Claus…'. Henrik believes that executive directors from other Greenpeace countries did not understand how important the argument that whaling is against international law was in Norway. This was because they lived in bigger countries that do not depend on international agreements as much as Norway does. They thought that this argument was too academic to be used in a campaign. But it is not at all too academic in Norway, the Askeladen country. Norwegians would understand that Norway should not ignore international agreements, 'particularly not with a high-profile figure like Brundtland in charge'. As Tor says: 'We're probably the best in class when it comes to following the regulations from Brussels. And we're not even a member of the European Union'.

But it would also be wrong to say that the reason for the failure of the whale campaign was that the international Greenpeacers were stupid. Nils repeats again and again that he can understand their point of view, their campaigning logic. John repeats again and again that he shared their view completely before he came to Sweden. Both emphasize that it took them a long time to arrive at their analysis and their new strategy. Nils and John don't phrase their analysis of the Greenpeace whale campaign in Norway in systems-theoretic terms. But they understood that all Greenpeacers involved did their best and did what they thought would win the campaign, based on their respective self-descriptions. Nils and John understood that people, and social systems, can fail to understand something because it is not information for them, because they lack the preconditions to identify and process it as information. They understood how difficult it is to metacommunicate. John and Nils both believe it was a lucky coincidence that the two of them were in the right place, at the right time, to get the opportunity to do what they did.

> *John*: We could understand what the Norwegians were thinking, we could understand what the Bambi-loving Brits were thinking. We, *we*, the pair of us, understood both sides. And took the time to understand both sides. Went through the effort, had the luck and chance to be able to do it. Together, we were able to come to the point, after a long period of time, to be able to interpret for the organization what this was about. And the fact that we could interpret for the organization was a major major thing…

This ability to interpret and translate combined with Nils' and John's abovementioned self-description (both on the individual level and on the level of Greenpeace as an organization) of People Persuaders. As has been described, the organizational self-description of Greenpeace as People Persuaders did exist in Greenpeace Norway, indeed in the whole Greenpeace organization, even before Nils and John arrived. But it was not always dominant.

> *Nils*: It's not just Greenpeace, it's very common in a campaign group that you make your plans based on your own world view. It is very unusual that campaign groups go out and try to understand, what do you think and how do you view this. What are you interested in and what can I do to make you interested. That is what I try to teach people now. A traditional campaigner paradigm, I believe, if you can talk about such a thing at all, is like, you know roughly what you want and then you take the tools you usually use. And then you drive hard, like, adamant.

Nils and John wanted Greenpeace to be a protest organization that takes their opponents' and audiences' world view into account when designing campaigns, rather than proceeding from its own world view.

The New Strategy

> *Alex*: There's a great story from Nils. What he says basically is, campaigning is easy. Any child can do it. Every child does it. All the time. Okay? And here's the example. Little Suzie wants an ice cream. She's defined the campaign goal, she wants an ice cream. She does a power analysis: Who is capable of producing the ice cream for me? Well, either Mum or Dad can buy it for me. She does a more in-depth analysis, saying, well, Mum will probably not buy it 'cause she'll just say, I bought you an ice cream yesterday. You can't have an ice cream every day. Dad is the one that I think I'll have an easier chance affecting. Okay. So, now she has to create a communication strategy. Now, she can't just go on to Dad and say, hey, give me an ice cream. She has to create an atmosphere, whereby Dad will want to buy her an ice cream. So, having done the analysis and having prepared a communication strategy, she goes to the roll-out. She goes up to her dad, she crawls up onto his lap, she hugs him, she says (*imitates a small girl's voice*) 'Daddy, you know you're never home. I miss you so much when you're gone. Last night I was crying myself to sleep when you weren't here. I love you so much Daddy. I miss you. Can I have an ice cream?'

Nils and John devised a new strategy for the whale campaign in Norway. What they identify as special about 'their' campaign strategy was that it comprised an external and an internal campaign at the same time. This is again proof of their ambition to be People Persuaders: they wanted to persuade both Greenpeace, as an organization, and the Norwegian public.

The goal of the new (external) strategy was to get Norway to stop whaling completely. The target group was 'the Oslo audience'. This is supposed to mean average Norwegians who are not, as relatives or neighbours of whalers, personally involved or interested in whaling, and who are therefore the disengaged majority who can bring about 'an atmosphere in which political change is unavoidable'. The message of the campaign was that whaling is against international agreements. What Nils and John wanted to emphasize was that this is a rational argument against whaling, not one born out of emotionality and sentimentality about whales.

> *Alex*: You cannot have too many messages. You can have one piece that you're communicating. John used this example of Coca-Cola. Coca-Cola is the best hangover cure that John has ever experienced. But Coca-Cola does not market itself as a hangover cure. What do you know when you drink Coke? You know that you're young, cool and American. And good-looking. That's the image. Not that your hangover gets cured. Now, Coca-Cola has put all of its global international power behind communicating this one message, and it works.
>
> Nils was saying stick to the message. That was something he was very strong on. Pick a message, you stick to the message. That is totally right. That is exactly how you should campaign. Don't make it complicated, and don't pick more than one, and stick to it. Every single press release, we won't say anything different, we say one thing. Illegal whaling, illegal whaling, illegal whaling, illegal whaling, illegal whaling. They come up with a different argument, you say illegal whaling. If they say, well, how's your mum feeling, you say illegal whaling.

What Alex says here fits very well with his categorization of Nils' and John's management of the Nordic Greenpeace organization as '*ursprünglich*'. Arguments with no more than ten words were what Greenpeace became famous for in the first place. It also relates back directly to John's view on what Norwegian campaigners did before 1999: produce confusion by using too many different arguments.

Nils and John also defined what was not the goal of the new strategy: to get Norwegians to like Greenpeace. Thanks to their focus group study, Nils and John were entirely clear that Norwegians had a negative attitude towards Greenpeace. They did not want to fight against that. Instead, they had the campaigners wear sweatshirts to press conferences in Norway which on the front said '*Drit i Greenpeace*' (Norwegian: Fuck Greenpeace) and on the back said '*... men redd hvalene*' (... but save the whales). They also had a Greenpeace folder produced in 1999 which was titled 'The Hairy Hippies' Guide to Norwegian Whaling'. It showed an old picture of – indeed very hairy and hippie – Greenpeacers of the early days, and explained the international agreements argument against whaling. Nils and John wanted to communicate that Norwegians were free to think Greenpeace a bunch of tree-hugging idiots, but they should listen to the rational argument against whaling. This is the clearest manifestation of Nils' and John's attitude that Greenpeace should be a People Persuader. If the goal is to Convince Other People, then you don't need to care about What Other People Think About You.

Instead of establishing Greenpeace as a credible actor in Norway, Nils and John wanted to create communicational space in Norway for someone else to take a stance against whaling. This followed the advice of the focus group study participants that a campaign against whaling should present arguments 'from an independent and trusted source'. Greenpeace was going to be the actor who again highlighted the issue, albeit in a cleverer way than before. But that was merely to draw attention to actors whom the Norwegian public regarded as rational, who spoke against whaling. Nils and John did not have any definitive plans as to who these rational actors would be. But they were optimistic that they existed and would eventually weigh in on the issue, once Greenpeace had created a renewed discussion and highlighted a rational argument against whaling.

Nils and John decided that the most important way for Greenpeace to highlight the whaling issue in Norway was to carry out direct actions. Again, this reflected their 'traditional' or '*ursprüngliche*' Greenpeace campaigning approach. Nils and John planned these direct actions very carefully: Greenpeace always had to act from boats that were smaller than the whaling boats. The big Greenpeace vessels were not to come within eyeshot of the Norwegian boats.

'Changed if You Do and Changed if You Don't'

To 'sell' their new strategy to their colleagues, John and Nils then conducted what they term their 'internal campaign'. After having analysed the campaign against Norwegian whaling so thoroughly, John and Nils understood that it would not be enough to merely point out that the campaign was flawed and suggest a different campaign strategy. That had been tried in the past, and it hadn't worked.

If an organization is caught in a pathogenic organizational double bind, attempts to change the situation that do not understand and address the (partly) unconscious organizational psychodynamics will usually fail. This is because they are likely to be incorporated into, interpreted according to, the pathogenic double bind pattern, and so will not resolve it (Hennestad 1990: 277f.). But even if a change agent has understood that the organization is caught in a pathogenic organizational double bind, it will not be easy to get the organization 'unstuck' (cf. Vince and Broussine 1996: 7).

This is because no matter how dysfunctional the consequences of the pathogenic organizational double bind may be, organizational members unconsciously bind themselves to it because it maintains and stabilizes the organization and protects organizational members from pain and anxiety. If a change agent attempts to remove the pathogenic organizational double bind, people will be faced with losing this protection and will resist (cf. Krantz 1999: 44; see also Diamond 1986; Papp 1980: 45). The organization is a classic case of an organization exhibiting 'resistance to change', even beneficial change. To effect real change, it is necessary for the pathogenic organizational double bind and the underlying problem to be 'worked through'.

The term 'working through' was first used by Freud (1978[1914]) when he sought to answer the question of why individual psychoanalysis takes a relatively long time: he argued that this is because the patient needs to work through his or her resistances. The concept has subsequently been subject to various interpretations (Aron 1991; Brenner 1987; Stewart 1963). From a relational perspective, Aron (1991) provides the following definition that seems compatible with different psychological and therapeutic approaches, as well as systems theory, and is also transferable to the level of a group or organization:

> [T]he working through process refers to the gradual collaborative task between analyst and analysand of transforming the analysand's inner representational world … constituted … by working models of relationships between the self and others. It is outdated and less adaptive models or schemata which are worked through and newer more highly adaptive relational models or schemata that are worked toward. (p. 103)

An organization caught in a pathogenic organizational double bind must analyse its working models of relationships and transform them into more adaptive or more clearly defined models of relationships. To achieve this, the resistances and defence need to be overcome, or given up (cf. Brenner 1987; Kets de Vries and Miller 1984: 155–166). Part of the process of working through a pathogenic organizational double bind is for the organization to learn how to metacommunicate: to communicate about its relational and communicational patterns in order to realize how these are inconsistent, contradictory and pathological. Just as individual 'working through' often requires the help of a psychotherapist, organizational 'working through' often requires the help of a skilled change agent (see for example Kahn 2004).

A particularly ingenious way of inducing metacommunication, reflection and 'working through' in a pathogenic double bind situation is the therapeutic double bind. This is a prescription characterized by the fact that one is 'changed if one does and changed if one doesn't' follow it (Watzlawick, Helmick Beavin and Jackson 1968). A therapeutic double bind is a special form of 'paradoxical intervention' or 'therapeutic paradox'. It is noted in the literature that paradoxical interventions, such as therapeutic double binds, are useful as a therapeutic strategy particularly where the pathological structure is long-lived and entrenched and does not seem to respond to other attempts at solution (Papp 1980: 45; Beutler, Moleiro and Talebi 2002: 214).

One example of a therapeutic double bind is Bandler and Grinder's (1975) story of a group therapist talking to a woman in the group who feels unable to ever deny someone's demand on her (to say no to someone). The woman ascribes this to a traumatic childhood experience where her denying someone's demand seemed to have very painful consequences. She is caught in a pathogenic double bind, because she feels bad if she says no to someone, but she also feels bad about not being able to say no and having to do everything that is asked of her. The therapist gives her the order to say no to each group member about something during the therapy session. The woman strongly refuses for several minutes. The therapist then points out that she has insistently said no to him and that this has not produced any negative consequences. The woman experiences this as a real eye-opener and immediately feels much stronger about saying no to people. This is a 'changed if you do and changed if you don't' prescription because, if the woman follows the therapist's prescription, she has successfully said no to someone; if she tells him that she will not do it, she has successfully said no to him (Bandler and Grinder 1975: 169ff.; see also Watzlawick 1978: 104f.).

Change agents may put an organization caught in a pathogenic organizational double bind in a therapeutic organizational double bind. A therapeutic organizational double bind implicitly recognizes the homeostatic function of the pathogenic organizational double bind. The change agents do not attempt to 'rationally' argue about or break the organization's defence and resistance to change, which would merely compound the 'stuckness' (cf. Smith and Berg 1987). Instead, they try to channel the power of the pathogenic binding in a constructive direction (cf. Beutler, Moleiro and Talebi 2002: 214; Shoham-Salomon, Avner and Neeman 1989).

A therapeutic organizational double bind provides an opening for the organization to learn new and better ways of relating than those it is used to. The woman in Bandler and Grinder's example is forced into an experience that contradicts her model of relationships, which implies that it is impossible to say no without grave consequences. Similarly, members of an organization in a therapeutic organizational double bind are forced into an experience which contradicts the model they have of the world (cf. Bandler and Grinder 1975: 169) and specifically of relationships. This enables metacommunication about this model and about the contradictory injunctions the organization is trying to follow, as well as working through of the organization's resistances to change and dysfunctional model of relationships. The woman in group therapy learns that it is possible in relationships to say no (see also the case reports in Hare-Mustin 1976 and Papp 1980). Likewise, a therapeutic organizational double bind makes new kinds of behaviour in relationships and more viable models of relationships accessible or plausible for the organizational members. To use Aron's words, these can then be worked towards. Because the therapeutic organizational double bind provides experience and practice of metacommunication and conscious change in behaviour, the organization's overall ability with regard to these may also be improved.

For a therapeutic organizational double bind to be successful, three conditions must be met that are analogous to the conditions for a successful therapeutic double bind in psychotherapy, e.g. family therapy.

Firstly, the change agents need to grasp the core of the organization's relational problem and the way in which the organization 'profits' from the pathogenic organizational double bind. But the change agents' understanding of the organization's pathogenic double bind situation must be 'meta' to that situation (cf. Watzlawick, Weakland and Fisch 1974; Wilder 1979: 176). The change agents must then formulate the therapeutic organizational double bind in such a way that it addresses the underlying problem (cf. Willke 1999: 131ff.; Olson 1972), yet signals to the organization that they refuse to be drawn into the pathogenic double bind logic (cf. Aron 1991; Kahn 2004). Getting drawn into such a logic is a very real danger for those trying to break it (Hennestad 1990: 277f.; cf. Kahn 2004). A successful therapeutic organizational double bind prevents the organization from interpreting the change initiative in the familiar pathogenic double bind pattern. The organizational change agents' task in formulating the therapeutic organizational double bind is similar to the task of the psychoanalyst, who must find a way to enter a patient's subjective world without being assimilated into an old object schema (see Aron 1991: 104). The change agents use their understanding of the pathogenic double bind logic to lead the organization in learning how to escape it (cf. Watzlawick, Helmick Beavin and Jackson 1968: 236–253). A therapeutic double bind seems to be a 'paradoxical' intervention only if seen within the context of the original, i.e. pathogenic interpretation of the situation (Watzlawick, Weakland and Fisch 1974).

Secondly, while the organization is unable to learn and is resistant to change, there must be a will to open up and 'learn how to learn' on the part of the organization. If the woman who cannot say no simply walks away from the group therapy when the therapist gives her the task of saying no, there cannot be any change. Similarly, the organization must admit that there is a problem that needs to be solved and commit at least to a degree to the therapeutic organizational double bind relationship with the change agents. Organizational 'pain' may be a motivation for such an attitude (Kets de Vries and Miller 1984: 166), although the change agents can also try and motivate the organization to embark on 'working through'. They can harness the organization's strengths for the 'working through' process (cf. Schein 2006: 299). They can also provide concrete opportunities for action through which the new ways of relating that the therapeutic bind makes accessible can be practiced (cf. Valenstein 1983) and offer an attractive vision of the future (cf. Kotter 1995: 63; Ford and Ford 1994).

For the organization to 'bind' itself to the therapeutic intervention (cf. Watzlawick, Weakland and Fisch 1974: 115), there must be trust between the organization and the change agents. 'Working through' can be quite painful and anxiety-provoking for organizational members; indeed, they may feel that their entire identity is threatened if this identity is strongly dependent on organizational membership. The change agents must therefore provide an 'organizational holding environment': they have to create an ambience that holds in place the identities of organizational members (see Van Buskirk and McGrath 1999: 806, 812). The concept of holding goes back to Winnicott (see for example Winnicott 1960), who understood it to mean parental care, or environmental provision, in the earliest stages of life (p. 589). Holding protects from disruptive, anxiety-provoking impingements, and enables the child to build up a sense of continuity of being (p. 594) and an identity in his/her own right (pp. 587–590). A psychotherapeutic setting can be a holding environment in that the therapist signals to the patients that s/he understands the patients' anxieties and makes it possible for the patients to experience these anxieties in a safe way, so that they are no longer as threatening to the patients' identities as before (Winnicott 1990 [1963]). In a therapeutic organizational double bind situation, organizational members must feel that it is safe to experience and communicate their emotions, and that they can reduce their defences against these emotions without being overwhelmed. This will support them in learning, growing and developing their identities (Kahn 2004: 8f.; Kahn 2001; see also Van Buskirk and McGrath 1999).

Nils and John had designed a new strategy for the whale campaign in Norway. In order to convince Greenpeace Nordic of it, they put Greenpeace Nordic in a therapeutic organizational double bind. They agreed with this interpretation when I presented it to them. John wrote in an email to me: 'I think the therapeutic double-blind [*sic*] idea is spot-on'. Nils wrote: 'Don't forget that I worked with therapeutic double binds for 10 years!' (my translation). Through their therapeutic double bind, Nils and John wanted to bind Greenpeace to the task of working through the pathogenic organizational double bind in Norway.

Their therapeutic injunction, or prescription, to Greenpeace Nordic was: 'Convince Norwegians that whaling should be stopped!' In the context of what has been said about the Greenpeace whale campaign in Norway, this injunction may certainly seem paradoxical: wasn't convincing Norwegians that whaling should be stopped exactly what had been impossible for at least fifteen years? With the help of this therapeutic organizational double bind, Nils and John wanted to sort out the relational problem that the Norwegian Greenpeacers had with their organization and that now had become Greenpeace Nordic's problem.

Their envisaged solution, their clarification of the relationship, was that Greenpeacers needed to 'persuade Norwegians' (injunction 1 of the pathogenic organizational double bind) that Norway should 'stop whaling' (injunction 2 of the pathogenic organizational double bind) by presenting a rational, convincing argument against whaling and not care if Norwegians disliked Greenpeace, so long as they heard the argument. The idea that Greenpeacers should be People Persuaders was thus (re)interpreted to mean not that Norwegians should support or be persuaded of Greenpeace, but support and be persuaded of the idea presented by Greenpeace. Conversely, the idea that Greenpeace should be an environmental outpost or frontline in the High North, holding out against the whale butchers with little support from the Norwegian environment, was (re)interpreted to mean that Greenpeacers should stand up bravely to attacks from their antagonists, but in such a way that it would convince average people of their case.

Nils' and John's prescription was a 'changed if you do follow it' prescription because, if Greenpeace Nordic wanted to follow it, it could not go on oscillating in its pathogenic organizational double bind, but needed to change the whale campaign strategy in such a way that Norwegians agreed with its argumentation. Consequently, Greenpeace Nordic would have to reflect on what it had done thus far, why it had not convinced Norwegians and how it could do better. Furthermore, the Nordic Greenpeacers would have to finally confront their non-Scandinavian colleagues. They would have to make absolutely clear to them, no matter what it took, that the whale campaign strategy had to change – which implied that the relationship between Nordic Greenpeacers and the rest of the organization would change as well.

The organizational reflection and discussion would have been therapeutic in itself. But if Greenpeace succeeded, the pathogenic organizational double bind would also have disappeared. If Norwegians became convinced that whaling was a bad idea, then they would stop whaling, and Greenpeace's anti-whaling campaign would not be a reason for them to dislike Greenpeace anymore. This path was the path that John and Nils' new whale campaign strategy projected.

On the other hand, the prescription was a 'changed if you don't follow it' prescription because, if Greenpeace Nordic did not want to follow it, then it had to acknowledge and discuss the reason why. How come it did not care about convincing the relevant public in this case, when for every single other campaign, Greenpeace relied on popular support? And after having this discussion internally, Nordic Greenpeacers would have to take it up with their colleagues outside Scandinavia.

Again, this kind of collective reflection would have been therapeutic. It would have clarified what Greenpeace wanted to be and do: Frontline or Persuaders? It would have brought the different emotions that different Greenpeacers had, both the emotional attachments to their self-descriptions and the mixed feelings about the whale campaign, out in the open, to be expressed and understood. It would have resolved the pathogenic organizational double bind – either because Greenpeace Nordic would have decided to follow Nils' and John's prescription and new strategy after all, or because Greenpeace Nordic would explicitly have given up its goal to 'gain support' and 'persuade people' in Norway, focusing instead on the single goal of stopping whaling (by whatever means necessary).

Nils and John implicitly recognized the stabilizing, defensive function of their colleagues' oscillation in the pathogenic organizational double bind. They did not try to impose the new strategy on the organization or to break the organizational resistance to change. Rather, their prescription attempted to channel the power of the double bind pattern into a constructive direction: a refusal to follow the prescription could have induced metacommunication and collective reflection just as well as an acceptance of the prescription.

Nils and John communicated their prescription in meetings and many informal discussions with their Greenpeace Nordic colleagues. They engaged in intense communication about the anti-whaling campaign in Norway with their Nordic colleagues. This communication included metacommunication about the communication that had been going on around that campaign previously. Their therapeutic organizational double bind provided Greenpeace Nordic with an experience that contradicted the dominant, pathogenic model of relating. John and Nils made it possible for Nordic Greenpeacers to access, and work towards, new models of relating.

As for the relationship between the Nordic Greenpeacers and their organization, John and Nils showed Greenpeace Nordic's employees a non-contradictory, consistent path of action that involved being both the High North Frontline and People Persuaders and promised support for it. This was a clear departure from the earlier situation, in which people had felt that their organization was putting them under pressure to follow irreconcilable injunctions.

John and Nils also did what they could to facilitate the type of discussion that would persuade Greenpeace International to accept their new strategy and to follow their example of consistency and supportiveness towards Greenpeace Nordic. They travelled extensively to meet and convince many international colleagues. Further, they held meetings in Stockholm and Oslo with representatives from offices around the world present. At these meetings they showed evidence that the campaigning that had been done in Norway had not been productive. They showed the picture that the High North Alliance had on their website of the big Greenpeace ship *Solo* chasing, and trying to spray water at, the little whaling boat *Senet*. They reported on the Askeladen folk tales, on the David against Goliath problem and on the problem of campaigning against each other. They presented their new strategy as the solution.

Nils and John both remember that their presentations seemed reasonably convincing to their international colleagues. The international Greenpeacers were surprised, but found it hard to contradict the arguments. But they were still sceptical, hesitant, grudging.

> *John*: I think there was one massively important thing that happened. It wasn't the presentation of the research that Nils and I had done. We said, here is what the research says, this is how Norwegians feel, and the international Greenpeacers didn't believe it, they didn't get it. Until we said, right, now all we want you to do is go out into the streets with a clipboard, here's a series of questions that we want you to ask average Norwegians, go and do some research as Greenpeace people. And so we got the Executive Directors of the organization and the campaign directors and the influential people in the main railway station in Oslo

talking to people for three hours, approaching people and asking them questions about Greenpeace, whaling and what they felt about it. And of course it matched; the questions produced the same answers as the research had. And that's when they got convinced. And they tried all the arguments they could think of on the Norwegians in Oslo Central Station, and then they tried our arguments, and they saw that the ones that we had said were the right arguments, were the right arguments.

And I truly believe that was the turning point for the campaign in Greenpeace. Major thing where things turned around, and left us two, three, four years of ability to do things we wanted and needed to do. At least the first couple of years after this turning point, we were able to change the way the organization spoke about whaling, to stop them saying silly things like whales' brains are bigger than our brains. And then have that available for Norwegians to use as they wanted to, on their websites.

The interaction of influential Greenpeacers with real, average Norwegians was a time-efficient, intense version of the 'clash of cultures' and 'clash of emotional worlds' experience John had gone through when he moved to Sweden from the U.K. It confronted the Greenpeace executives with the fact that Norwegians did not accept at all the version of reality they had always taken for granted. It exposed them to people who questioned Greenpeace's self-descriptions, Greenpeace's arguments, everything Greenpeacers thought was obvious and logical. Importantly, it exposed the Greenpeace bosses to Norwegians' emotions. This made the 'clash of cultures' experience much more impressive and convincing for the Greenpeace executives.[2]

This helped to make John's and Nils' analysis and suggestions understandable in the social system Greenpeace. It helped to prompt the reflection that John and Nils wanted to induce through their therapeutic organizational double bind intervention. At a later point John again emphasizes his belief that if 'we had just handed over the research to the international Greenpeace executives, it would never have worked'.

John: It can't be overestimated how important the fact that the research was confirmed by then talking to people in Oslo Central Station was. We said, you are going to fly up to Norway. You're going to come, this is important, and now you gonna go out in the street and you gonna talk to people. They didn't want to do that at all…

The new strategy required international Greenpeacers to change their way of working in some ways. For example, they were not allowed to use any arguments other than the one prescribed, anywhere in the world. They did not necessarily like that at first. But there were also positive incentives for the international Greenpeace organization to accept John and Nils' new strategy. It made the relationship with Greenpeace Nordic more constructive for them. International Greenpeacers no longer felt that they were being 'kept out of Norway' by their defensive Norwegian colleagues. The new approach showed that the Nordic office wanted to act on the whaling issue and wanted the international Greenpeace organization to do the same. In fact, Greenpeace Nordic wanted to carry out 'traditional' Greenpeace direct actions against Norwegian whaling vessels (see the next section for descriptions). The non-Nordic Greenpeacers appreciated these things.

> *Nils*: I mean it's obvious it was hard, because they had their way of doing things. At the same time it was easy. They weren't stupid, they said also that this is, hmm, okay, a completely new way to attack the problem, now we've been driving it in other ways for ten years, let's give it a chance, was what many were feeling. But then also, we invited the other countries: come here, take part. What Greenpeace Norway had done before was like, to keep them away all the time.

The international Greenpeacers ultimately accepted John's and Nils' new strategy, at least for a while. This meant that the relational problem between Nordic Greenpeacers and their organization was solved at least temporarily. As a consequence of the fact that Greenpeace Nordic's pathogenic organizational double bind and the underlying relational problem were tackled in this way, a new, well-defined and unambiguous way of relating to Norwegians also became possible for Greenpeacers. This was, of course, to present one convincing rational argument against whaling without caring whether Norwegians disliked Greenpeace, as long as the argument got heard.

Nils' and John's therapeutic organizational double bind addressed the core of the Norwegian Greenpeacers' relational trouble. Their therapeutic injunction clarified both the relationship between the Nordic Greenpeacers and their organization and the relationship between Greenpeacers and average Norwegians. However, it also communicated the fact that while Nils and John understood the Norwegian predicament, they would not be drawn into the logic of the pathogenic organizational double bind. Their interpretation of the situation of Greenpeace in Norway was 'meta' to that situation. Indeed, their therapeutic injunction 'Convince Norwegians that whaling should be stopped!' seems illogical or paradoxical only within the logic of the original pathogenic binding. Outside that logic, using confrontational stunts and communication to persuade average people to support an idea has always been Greenpeace's strategy and is not all that paradoxical for a Greenpeacer.

Nils and John were able to put Greenpeace Nordic in a therapeutic organizational double bind because many in Greenpeace Nordic were fed up with being 'damned whatever they did'. They were ready to accept real change if it was offered to them, even if it came at a cost.

> *Caroline*: I think there was probably a feeling that there were a lot of other campaigns going on. And a lot of other work to be done. And that, every year we were being asked to do something for Greenpeace International, that felt like it was taking resources away from other campaigns that we wanted to work with. And so, I think maybe that was what triggered the okay, then at least we'll have the control of what's going on. We'll campaign against Norwegian whaling again this summer, we'll do it again next summer, we're gonna continue working on this, but we're gonna do it our way.

Nils and John motivated Greenpeace Nordic to embark on a 'working through' process in the following ways. The fact that they deliberately made direct actions a core part of their new anti-whaling campaign strategy meant that they were harnessing a core tactic and core strength of Greenpeace. The direct actions provided perfect opportunities for Nordic Greenpeacers to relate differently to average Norwegians by putting out the 'Fuck Greenpeace…' message. As mentioned above, they were also an opportunity to relate differently to the international Greenpeace organization. Previously, some Norwegian Greenpeacers had attempted to prevent direct actions against whaling in Norway, but now international Greenpeacers were explicitly asked to come to Norway to help with these actions. Sure enough, my interviewees report instances of 'working through' that occurred during direct actions (see the quote by Caroline below). Nils and John's strategy included a clear and attractive projected outcome, with Nils and John monitoring and communicating within the organization the strategy's success – above all a change in media reporting (see below).

Another reason why Nils and John were able to put Greenpeace Nordic in a therapeutic organizational double bind was that they were able to provide a 'holding environment' for Greenpeace Nordic, which had not existed in Greenpeace Norway. They managed to create a comparatively safe situation in which Greenpeacers were willing and able to accept some discomfort in exchange for a real development. People working for Greenpeace Nordic trusted and respected Nils and John personally and trusted that they would invest the necessary time, effort and patience to 'sort out the anti-whaling campaign'. Alex has already been quoted in Chapter 2 as saying: 'Nils was really a leader. Nils comes in, in 1996, and he is the guru building it up. We all look up to him, he is the one to tell us what Greenpeace is, what campaigning is, and we all kind of like go "fearless leader, we go" and we do it'. Also, Greenpeace Nordic was financially self-sufficient and more autonomous than Greenpeace Norway had been, and Nils and John convinced international Greenpeacers to support their new campaign strategy.

An important part of the organizational 'holding environment' was that Nils and John intuitively understood the identity problems involved in the pathogenic organizational double bind. They understood that individuals and organizations are emotionally attached to their self-descriptions and find it hard to give up on these if they do not get equivalent ones in return. They helped by showing how their Greenpeace Nordic colleagues could be both People Persuaders and The High North Frontline, offering them organizational self-descriptions that were functionally equivalent to their old ones. (The literature on organizations supports the idea that organizational members will resist change insofar as it threatens a functioning organizational self-description; see Brown and Starkey 2000; Eilam and Shamir 2005; Vince and Broussine 1996. In the type of organizational context described here, the importance of organizational and individual identities and self-images that are stable yet capable of developing is also stressed by Schein 2006: 296.)

If they followed John and Nils' strategy, Nordic Greenpeacers would be able to describe themselves as 'those who are rational and confront a destructive thing'. The destructive thing was whaling, of course, but they also confronted a campaign that had been destructive for Greenpeace internally. Nils and John communicated to their Nordic colleagues that they should be People Persuaders who both persuade Norwegians to adhere to international agreements and persuade their international Greenpeace colleagues of the new strategy. And Nils and John communicated further to their Nordic colleagues that they should be The High North Frontliners who stand up bravely to opposition both from Norwegians and from other Greenpeacers, in order to persuade everybody of their arguments in the end. This meant that Greenpeace Nordic had a rational and grown-up sort-outer role both with regard to Norway and Greenpeace internally. In return for this self-description, Nordic Greenpeacers were willing to invest time and energy in a campaign they did not like very much. John and Nils coaxed Greenpeacers into accepting certain inconveniences by offering them emotional compensation, changed self-descriptions that satisfied them.

Eventually, both Nordic and international Greenpeacers understood Nils' and John's messages, at least to a degree, and showed a will to implement the strategy it had taken Nils and John so long to work out.

Effects of the New Strategy and the Therapeutic Organizational Double Bind

In 1999, Greenpeace implemented a 'big push' in the whale campaign in Norway, using the new strategy, and running direct actions which received quite extensive media coverage.

In May, Greenpeace Norway informed the public that Greenpeace's *Rainbow Warrior* would be sailing in the North Sea to campaign against whaling (Mathisen 1999). Greenpeace then undertook a campaign activity which Nils calls 'fucking brilliant, incredibly good'. This was to prewrite letters to Norwegians in English and Norwegian and send one of these to each of their German and Swedish supporting members with a cover letter asking them to sign it and send it to a Norwegian whose address (picked from the Norwegian telephone register) was also included. A press release from Norsk Telegrambyrå AS about this reports that 'Greenpeace has asked its 530,000 German supporting members to sign an English letter (they have read a German translation) and send it to a Norwegian who is randomly picked from the telephone book'. The press release mentions that the main argument in the letter is that Norway should stop whaling out of respect for international conventions and that the Norwegian hunt constitutes a bad example for other nations, and that 'the tone in the letter is personal' (Veigård 1999, my translation).

A Swedish Greenpeace *aktivitetshäfte*[3] dating from 1999 contains a long letter in Norwegian and an address to a Norwegian. Supporters are asked to sign the letter and send it to 'their' Norwegian – 'if you don't send it, that Norwegian won't get to know the arguments'. The letter begins by detailing the Greenpeace opinion that Norwegian whaling is against international agreements (see above), and explains that to accept commercial whaling could lead to renewed overexploitation. It goes on to point out that illegal whale products (containing for example blue whale) have already been found on the Japanese market; that there is a surplus of whale blubber in Norwegian storage, some of it more than ten years old; that many countries traditionally participated in whaling but stopped when it became clear that many whale species were in danger of extinction; and that it is more profitable to offer whale safaris than to hunt whales. The letter ends with the words:

> I know there are cultural differences between our countries, and I can understand that some of the arguments against whaling can seem exaggeratedly emotional. But some arguments overshadow [*sic*] the cultural differences. I think most Norwegians think it is wrong not to care about international agreements, and that it is wrong to put

> endangered animal species at risk by lobbying for [lifting] export [restrictions]. I also think most agree that it is also wrong to build up storage of food no one wants to eat. And that all this is especially wrong when there is a solution which would serve local communities better. Thank you for taking the time to read this letter. You live in a fantastic country, and I think it's a pity that whaling should destroy that country's good name and reputation. Best regards, ... (my translation; see also Gjørvad 1999)

Nils says he thought this was 'one of the absolutely best campaign activities' because 'it was so clearly something different than what people associate Greenpeace with'. This was not a big, money-hungry foreign organization putting pressure on Norway. This was a 'completely normal person', a human being, from somewhere in Germany or Sweden explaining in a personal letter that they were opposed to whaling.

> *Nils*: I don't believe we could find out exactly how many letters arrived, but we did a non-scientific test. We asked everybody we met and we asked all the volunteers of Greenpeace Norway and so on, and it turned out that everyone we talked to had either got a letter themselves or knew someone who had received such a letter.

On 12 June 1999, a British Greenpeace activist was seriously injured when a Coast Guard boat and a Greenpeace dinghy collided. He was taken to the hospital in Stavanger by rescue helicopter. On 13 June 1999, Greenpeace conducted a direct action against the whaling vessel *Villduen*. These events took place outside Lista in Southern Norway (Mikalsen 1999a; Westengen 1999; N. N. 1999a). On 16 June 1999, *Dagbladet* reported that Greenpeacers had been invited to have some whale pizza in Borhaug, Lista (but had not accepted the invitation) (Mikalsen and Schjølberg Ulshagen 1999).

On 8 July 1999, *Dagbladet* and *Dagens Næringsliv* reported that the British newspaper *The Independent* had criticized Norway for its whale blubber stocks. The *Independent* had highlighted the fact that Norway defied the international whaling moratorium (N. N. 1999b; Gjørvad 1999).

Dagbladet reported on 11 July 1999 that whales had been shot illegally after the introduction of the catch ban on 1 July. (The ban was introduced by Norges Råfisklag, the Norwegian Fishers' Union, because the storehouses were full and wholesalers could not take in any more.) This was seen as damaging for the whaling industry's international reputation and thus potentially as 'a present to Greenpeace, Paul Watson and the public opinion'.

The article shows the 'Fuck Greenpeace, but...' tendency the campaign was designed to provoke. Greenpeace is portrayed as a group of 'whale friends' who don't understand Norwegian whaling, and named in the same breath as Paul Watson, but whaling is critically discussed as well (Tonstad 1999; my translation).

On 12 July 1999, Greenpeace conducted a direct action against the whaling vessel *Kato* outside of Egersund, Southern Norway. One of the whalers fired a rifle shot, which hit one of Greenpeace's rubber dinghies. According to the whalers, he was trying to shoot at a whale which had been harpooned in order to kill it, as he was obliged to do. The whaler was later given a six-month suspended sentence and a fine of 20,000 Norwegian kroner for shooting a hole into the rubber dinghy with Greenpeace demonstrators on board. The court also ruled that he had to pay 6,000 kroner for costs of the court case and have the rifle in question taken away from him (Skjalg 1999; 2000). The whaler decided not to appeal the sentence (N. N. 2001).

Member of Parliament and pro-whaling activist Steinar Bastesen was a member of the crew onboard *Kato* when the shooting incident occurred. He was quoted as saying 'We did not shoot to hit the rubber boat, but if Greenpeace are so stupid as to put themselves between us and a hurt whale, they have to be aware of the risk. It must have been a ricochet that punctured their boat' (Mikalsen 1999b, my translation; see also Olaussen 1999).

On 15 July 1999, *Dagens Næringsliv* ran a leader on page 2 which referred to the shooting incident. The article compared the profit from Norwegian whaling to the expenditures on research and diplomatic activity about it. The conclusion was as follows: 'Should the hunt continue? We don't believe so. The only argument is, as far as we can see, that it's not fun to give in because of Greenpeace actions' (N. N. 1999c, my translation).

This is exactly what Nils and John had been hoping for: a reputable Norwegian 'third voice' coming to a conclusion of the 'Fuck Greenpeace, but note the rational argument against whaling' type.

On 17 July 1999, *Aftenposten* reported that the skipper of the *Kato* had invited Greenpeace's campaign leader by letter to come on board the *Kato* to talk. The campaign leader said that he appreciated the initiative, but politely declined. He said that he hoped to meet the whalers at a later date (N. N. 1999d).

The Norwegian whale campaigner published debate articles in *Dagens Næringsliv* on 27 July and 23 August 1999. He explained that Greenpeace, when protesting against commercial whaling and the export of whale products, did not view the Norwegian hunt as an isolated issue, but saw a risk for increased hunt by other nations, also on really endangered whale species. He emphasized the importance of following international agreements (Pleym 1999a, 1999b). He also wrote a Letter to the Editor that was published in *Nordlys* on 25 August 1999, titled 'Norway Ignores IWC', presenting the same arguments (Pleym 1999c).

In addition to provoking Norwegian media coverage of the whale campaign of the type which Nils and John had envisaged, the therapeutic organizational double bind induced some working through within Greenpeace Nordic. As mentioned above, Nils and John engaged in intense metacommunication with their Nordic colleagues about the communication that had been going on around the anti-whaling campaign. And the Nordic Greenpeacers in turn engaged in discussions with their international colleagues, arguing for the new strategy and for the idea that Greenpeacers should be People Persuaders, as well as the High North Frontline stopping the bloody whale butchers.

> *Caroline*: I've been on a ship that was preparing for some actions in Norway, and a lot of the crew members were from the U.K., and we were discussing how we were going to do the actions. The crew was very intent on stopping whales from being killed. Every whale that they saved, felt like a success. And our point of view was that this isn't sustainable for us as a campaign because if we do this every summer with all our resources, maybe we'll save 40 whales a summer and then what? You know, we have to get the debate up and then get a stop on whaling, then we've won. In the process, maybe whales will be killed. I think there is a willingness within Greenpeace to go and to stop things. And that's a very basic like, activist way of thinking. You can go save your twelve whales if you want.
> *Juliane*: But if you fuck up the public opinion in Norway completely…
> *Caroline*: …then you'll have to come back here and save, you know, your twelve whales a year, but you haven't won, you know, wrong term, you haven't won anything.

The following is a judgment by a Nordic Greenpeacer made in 2005 which suggests that John and Nils' intervention had a therapeutic effect on Greenpeace Nordic.

> *Alex*: I think, in the end, the new approach was a success 'cause we'd been so afraid in Greenpeace Nordic to do anything about whaling. Everything Greenpeace does in Norway now, has like, 'whaling whaling whaling' echoing somewhere in their subconscious. It's like, this campaign we just did now? In the Barents? That was a tremendous success in the media. We're like, public heroes. And when you listen to

the Norwegian campaigners talk about it like, aah, this is great, finally, we're turning over this leaf of wha... you know, it's like, in Norway everything goes back to whaling. So, it was good to confront it. You should confront your fears, you know. It's like if you're afraid of spiders, you should pick some up.

Ultimately, however, the new approach's success must have been limited simply because it only lasted one year.

Notes

1. In a Greenpeace-internal paper I was given by a Greenpeace lawyer the argument is presented in more detail as follows:

> 'Art. 65 of the United Nations Convention on the Law of the Sea (UNCLOS) provides that in the case of cetaceans, States shall work through the appropriate international organizations for their conservation, management and study.
>
> The "appropriate international organization" in this case is the International Whaling Commission (IWC).
>
> The IWC has passed a moratorium on whaling. The IWC has also called upon Norway to cease whaling in 1996, 1997 and 1998. Norway has refused to do so. ...
>
> By first objecting to the IWC's moratorium and then by resuming its whale hunt, Norway not only defies the will of the international community ... but fails in its duty under Art. 65 UNCLOS to work through the appropriate international organization for the conservation and management of marine mammals and to co-operate with a view to the conservation of marine mammals. Norway's actions including its decision to resume whaling despite the IWC's moratorium is [*sic*] therefore in contravention of art. 65 UNCLOS.
>
> ... Norway based its objection to the moratorium on art. V (3) (c) of the 1946 Whaling Convention. This article provides that when future amendments are made, States will have the option to object to them. However, since the moratorium was passed, the IWC has repeatedly condemned Norway for its resumed whaling and called upon Norway to respect the moratorium.
>
> The 1982 United Nations Convention on the Law of the Sea was ratified by Norway on 24 June 1996 and has been in force since 24 September 1996. Both UNCLOS and the 1946 Whaling Convention are binding international agreements which supercede Norwegian Law. Norway cannot claim to be working with the appropriate international organization when it continues to practice whaling despite both the international moratorium and the call of the IWC to stop whaling'.

Greenpeace used this argumentation publicly, as documented in a 1999 brochure for Swedish supporting members, which will be discussed in more detail in the section on 'Effects of the New Strategy'.

2. These emotions were not only communicated to them, but also became perceivable to them in Norwegians' habitus. Because perception of emotions via habitus is 'fuller', more immediate, than communication about emotions without perception of habitus (see the introduction), this contributed to making the 'clash of cultures' experience convincing for the Greenpeace executives.
3. 'Activity brochure', a small brochure Greenpeace regularly sent out to its Swedish supporters, containing various activities for supporters to help campaigns, such as sending postcards to politicians, etc.

Chapter 5

'From Direct Actions to Dialogue'

Greenpeace Campaigning in Norway from 2000 Onwards

John and Nils both resigned from their jobs with Greenpeace Nordic to take up positions with other organizations in 2000. With the two protagonists of the 1999 whale campaign strategy gone, strategy changed again.

The Campaign Strategy since 2000

Strategy in the campaign against Norwegian whaling since 2000, as explained by my interviewees during my interviews with them in 2005 and 2006, can be summarized as follows. Whaling has been put into the broader context of the oceans campaign. Thus, there is no such thing as 'a whale campaign' anymore. Communication is to promote a 'holistic view' (Fredrik) on the anthropogenic degradation of the oceans, including issues ranging from climate change and pollution to overfishing and whaling. Whaling is not to be too big a part of the Greenpeace communication in Norway. Greenpeace uses arguments against whaling that it believes Norwegians perceive as 'rational' and convincing, in particular, the argument that Norwegian whaling is not in line with international agreements (cf. Helle and Stenerud 2003). With regard to this point, strategy from 2000 onwards has clearly built on Nils' and John's work. Greenpeace's strategy is dialogue-oriented, meaning that confrontational methods like direct actions are not used against Norwegian whaling, although Greenpeace Nordic does use confrontational methods when campaigning about issues other than whaling. According to interviewees, Greenpeace Nordic since 2000 has also worked with the International Whaling Commission and worked to prevent the export of whale products from Norway (see also Pleym 2000; but see High North Alliance 2002).

> *Juliane*: So if you should summarize the strategy change in the whale campaign during the time you worked for Greenpeace…?
> *Lars*: From direct actions to dialogue. It really is that simple to summarize. And it was controversial back then in 2000, but Greenpeace Nordic were agreed on it. We said, we have run our forehead into walls till it bleeds, in Norway, and what we have done has really made Norwegians more resistant. GPI said, OK, we can accept not doing hard-hitting actions. And we could show later, that the public attitude towards whaling and Greenpeace had changed in Norway, not much, but it had changed.

In 2000, this strategy was implemented in the form of a Greenpeace ship tour to coastal communities in Norway. The ship crew did not conduct direct actions, but invited people in fifteen Norwegian ports to starting a dialogue, trying to explain the Greenpeace arguments against whaling and Greenpeace work more generally (see also N. N. 2000; Emberland 2000; Glosemeyer 2000). Greenpeace also organized more letter-writing in 2000, asking Dutchmen to write to Norwegians to protest against commercial whaling and the export of whale products (Andersson 2000).

> *Juliane*: Did other Greenpeace offices accept the port travel dialogue strategy?
> *Fredrik*: The 2000 approach it was some scepticism against, because it was a perception of, how can Greenpeace be going to Norwegian ports while there are whales being killed out there. I mean, they feared reactions from their supporters. But as this ship tour during 2000 developed, and you had a lot of positive reactions from Norwegian press, you had the Norwegian prime minister visiting the Greenpeace ship, people saw that it was a good approach in terms of combating prejudices, communicating the rational argument in Norway. It probably didn't help to promote the campaign, increase public opposition to whaling, outside Norway. But it helped the campaign in Norway. Which, by then, all Greenpeace offices fully agreed was an important goal.

In the following years, Greenpeace focused on running a fisheries campaign in Norway. It is considered important to 'take sides' (Björn) with coastal fisheries by addressing issues such as tanker traffic, showing that Greenpeace has a genuine interest for the fisheries branch of the economy to function and not to suffer from environmental destruction (cf. Hernes and Mikalsen 2002). If this is the context, Greenpeacers believe they can also get the whale message out.

Greenpeace Nordic does not want to draw attention to the whale issue in Norway. It opposes the Norwegian whale hunt, but wants to scale the issue down, not raise it high on the agenda. This is because Nordic Greenpeacers see Norwegian whaling as a small question in the big scheme of things. This is a view with a long-standing tradition in the Scandinavian Greenpeace offices, and one not without merit if one compares the environmental consequences of the Norwegian minke whale hunt to those of climate change, for example. Nordic Greenpeacers also believe they will save more whales in Norway this way than by putting the question in the limelight the way Greenpeace did during the 'whale wars'. Alex says: 'Basically, those strategies, communicating with coastal communities, communicating that whaling is illegal, and working politically at IWC, are geared at nonexpansion. Whereas in reality what you're waiting for is for whaling to just die, because there's no market'.

Most interviewees judge the strategy since 2000 to have been successful. They don't declare a win, because Norway is still hunting. But they see the advances in the whale campaign as good and they are certain that a different, non-dialogue-oriented strategy, particularly one using direct actions against Norwegian whaling vessels, would have made these advances impossible or even done additional harm. They argue that the demand for whale meat and the political interest in the question have fallen, that whalers are catching below the allowed quotas and that the export of whale products from Norway has been stopped.

> *Sander*: I think that the low key strategy on whaling, and the high focus on other issues, is helping the environment a lot. And it's also advanced the whale campaign, but in a much more sophisticated way. And we can see it on a number of factors, that the consumption is going down, the attention for the issue is going down, the tension between us and the coastal communities, including the whalers, is going down, down, down. And it's gonna basically only remain a few very old-fashioned individuals that still insist on whaling as a very important thing.

> *Fredrik:* The new approach did indeed lead to a couple of things which we did not achieve while we had the old campaign. Stopping export, slowly but surely changing the Norwegian public. Being able, as a part of the latter, to work on other issues with the population in Norway, on fisheries, on oil tanker traffic close to the shore, etc. That would not have been possible before '99. I think what's hard for Greenpeace to realize is that it might take time to end whaling completely. Even if we have to continue to fight and be proactive, at some level we have to

realize that some campaigns do take time. And you do experience losses in the process.

In 2014, the Norwegian Broadcasting Corporation NRK published the following article, quoting the leader of Greenpeace Norway:

> Greenpeace: – The [Norwegian] whaling business will die out on its own. – Globally, the fight against whaling is basically won. The few countries that still hunt whales take small numbers, says Greenpeace Norway's executive director, Truls Gulowsen. … – The plans to [establish] oil operations in the Lofoten islands, on Bjørnøya and … in the Arctic are much bigger and more important issues to use limited environmental [activist] resources on, than what 20 or so whaling boats do during the summer, thinks Truls Gulowsen … . Greenpeace nonetheless thinks that Norwegian whaling should be stopped as long as it contravenes international agreements. – Many still have strong feelings about whaling, but I hope our [direct] action in 1999 will be the last [against Norwegian whaling] that Greenpeace is involved in in Norwegian waters, [Gulowsen] says. (Lysvold 2014, my translation)

Statements from a Greenpeacer such as 'Other things are more important than what 20 whaling boats do during the summer' or 'I hope we won't do direct actions against Norwegian whaling any more' would have been unthinkable in 1994, at the height of the 'whale wars'. Any Norwegian Greenpeacer who had said similar things to journalists in 1994 would in all likelihood have been fired. It seems that the Norwegian 'whale wars' really are over for good.

However, the positive judgment about the strategy since 2000 can be questioned. It is probably true that political interest in the whaling issue in Norway has decreased, thus reducing the status of the issue as a symbol of national autonomy, etc. It is also correct that whalers have often caught fewer whales than the quota allowed (Statistics Norway 2010; Norwegian Ministry of Fisheries and Coastal Affairs 2005, 2006, 2008; Andersen and Indsetviken 2014). However, the Norwegian government raised the allowed quota from 797 minke whales in 2005 to 1,052 minke whales in 2006, and quotas have remained comparatively high since then (1,286 minke whales in 2010, 2011, 2012, 2013 and 2014) (Norwegian Ministry of Fisheries and Coastal Affairs 2004, 2005, 2010, 2013; Directory of Fisheries 2014). Further, Iceland resumed whaling for scientific purposes in 2003 and commercial whaling in 2006 (Icelandic Ministry of Fisheries and Agriculture 2006, 2009), and haggling within the IWC shows no signs of becoming less fierce (Vidal and Adam 2010; Vaughan 2010; Vidal 2011). Thus, there is still pressure in favour of whaling on the international level. Will Norwegian whaling really just die out? Former Greenpeace campaigner Hans, for one, is not very positive in his judgment of the whale campaign: 'Norway is increasing the quota every year. Greenpeace has effectively lost this campaign. If I was in the Norwegian Foreign Office, I would say that well, we've gotten it through, and they've gotten it through because they have argued the science and the numbers'.

It is always difficult to judge whether a protest group's activities have been successful, not least because they interact with so many other events which the group does not control (cf. Giugni 1998). The reason why it is, ultimately, especially hard to judge whether current strategy in this particular campaign has been successful or not is that there is no longer only one, single, clear campaign goal. In 1999, the single goal of the strategy designed by John and Nils was made extremely clear: Norway must stop whaling completely. Judging from that, campaigning from 1999 up to now has been just as unsuccessful as campaigning during the 1980s and 1990s.

But that is no longer the only yardstick Greenpeace Nordic uses to measure its success. While Greenpeace Nordic still opposes the Norwegian whale hunt, it has also (again) been a goal since 2000 to show Norwegians that 'Greenpeace is more than whales'. Greenpeace Nordic wants to become accepted as a valid spokesperson for a legitimate set of interests (see Gamson 1975) such as protecting the Arctic and working against climate change, in Norway. Its employees are convinced that campaigning to stop whaling of the type that was done in 1999 would make this impossible. (Apart from that, they are also convinced that this kind of campaigning to stop whaling would make stopping whaling impossible; see also Lysvold 2014.) Greenpeace Nordic wants to reduce tension between it and Norwegians. It wants to be able to cooperate with Norwegians.

> *Sander*: We are there now to fix environmental problems like we do all over the world. And still be Greenpeace, with all our history and all our positions, including those on whaling.
>
> We believe we can see quite clearly that people, that earlier wouldn't talk to Greenpeace about anything, are now supporting us with evidence about where we can find the coral reefs, which politicians we need to pressure to achieve… so the noncommunication has disappeared, even with whalers.
>
> I think it's always a good thing to achieve dialogue and communication, on a reasonable level. And also to show that this organization is not only about whales, but it's about protecting the general environment. And we believe also that that approach will in the

future create much less space for the very hardcore, nationalistic whaling position of Norway.

This positive and optimistic judgment about being able to achieve support and cooperation in Norway with the new strategy and being able to work successfully on other issues than whaling is shared by a number of both current and former Greenpeacers.

But even the positive judgment with regard to these goals can be questioned. According to Greenpeace Nordic's annual report of 2013, 6,734 individuals in Norway gave money to Greenpeace in 2013, compared to 26,266 in Denmark and 98,218 in Sweden. It is certainly true that an organization can have a prominent influence on policies in a country even with a small number of supporters. It is surely also true that it would be unrealistic to try and achieve an enormous increase in Greenpeace Norway's supporter figures, given the history. And there is no reason to doubt current Greenpeacers' reports about more trustful and respectful interaction with people who earlier despised Greenpeace.

But still, do these figures justify a judgment that Greenpeace in Norway today is a supported organization, an organization backed up by the people, an organization which represents what people care about? Has the strategy of scaling down the whaling issue, of being non-confrontational on the issue and focusing on other issues, thus been successful, in the sense of making Greenpeace a valid spokesperson for legitimate interests? Former Greenpeacer Søren says that while he believes Greenpeace's supporting membership potential in Norway to be limited because of historical events, he feels that Greenpeace has not yet taken full advantage of this potential. Ex-Greenpeacer Hans voices his disappointment about the fact that 'Norway today doesn't have a good solid environmental organization and hasn't had one for years'. I think that the reason why Greenpeace Nordic sticks to its strategy is not so much that the strategy has been successful, be it in the sense of stopping whaling or in the sense of gaining support among Norwegians. I think that the fact that Greenpeace Nordic sticks to its strategy can be explained by the power of pathogenic double bind patterns.

Relapse into the Pathogenic Organizational Double Bind

The current strategy in the whale campaign is a relapse into the old pathogenic organizational double bind, albeit in a mitigated form. Greenpeace is again trying to follow two injunctions in Norway that contradict each other. On the one hand, it is attempting to gain support in Norway, and its work on issues other than whaling is, in principle, certainly apt to reach that goal. On the other hand, Greenpeace has not abandoned the organizational goal of stopping Norwegian whaling. The fact that supporting member figures in Norway continue to be low seems to indicate that Norwegians still see a contradiction here. Greenpeace is simply trying to make this contradiction less acute.

> *Nils*: I believe that had Greenpeace Nordic continued this campaign using John's and my strategy, they could have been successful.
> *Juliane*: Why didn't they continue?
> *Nils*: I believe they believed they continued (*chuckles*), but they didn't. It was a little that they tried to save Greenpeace and save the whales I think. And with that, they left the strategy we had.

A therapeutic organizational double bind does not make a pathogenic organizational double bind disappear quickly. (Similarly, the woman who is 'forced' by the therapist to say no to someone will for some time continue to find it difficult to say no to people.) Instead, the two organizational double binds will continue to exist alongside each other. The therapeutic organizational double bind, if not abandoned, will ensure that a 'working through' process takes place, through which the pathogenic organizational double bind is resolved. The change agents need to follow through with the

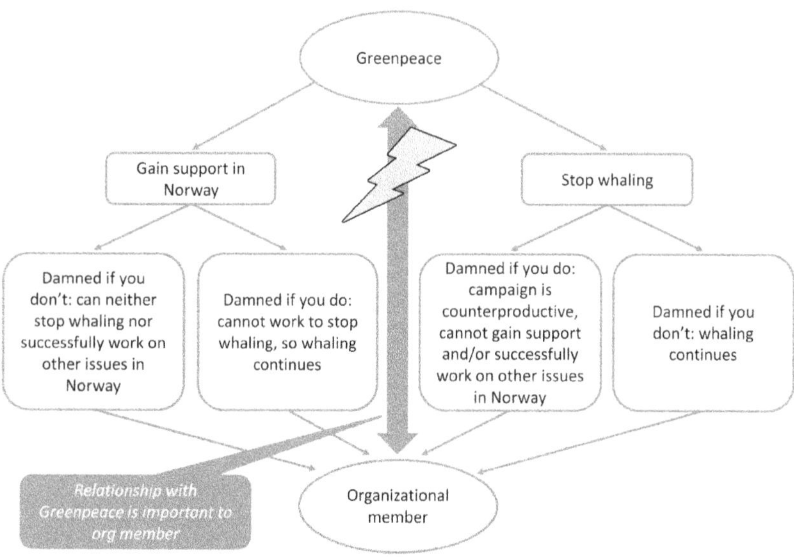

Figure 5.1. Greenpeace Nordic's new pathogenic organizational double bind

binding for a sufficient period of time, because it can be easier for the organization to fall back on pathogenic double bind patterns than to resolve its relational problems (cf. Papp 1980). The organization needs to understand that 'the only way out (of the pathogenic organizational double bind) is (working) through'. The therapeutic organizational double bind needs to be maintained long enough for the organization to properly grow out of its pathogenic organizational double bind.

In the case of Greenpeace Nordic, it would have taken several years of people like Nils and John communicating the therapeutic prescription and the new strategy for these to become truly ingrained in Greenpeace Nordic's ways. It would also have taken years for the changed message Greenpeace was sending out to really reach and be understood by the Norwegian audience, thereby making Norwegians less sure about whaling and less disapproving of Greenpeace and thus slowly resolving the pathogenic organizational double bind.

John and Nils bound Greenpeace Nordic to their therapeutic prescription. Because Nordic Greenpeacers trusted and respected them, and because they created a 'holding environment' for Greenpeace Nordic that made 'working through' with the help of the therapeutic organizational double bind bearable, Nordic Greenpeacers accepted this binding. However, after a brief period of organizational reflection, John and Nils left, the therapeutic organizational double bind disappeared, and so the organization slipped back into the pathogenic organizational double bind. The very 'holding environment' Nils and John had created made it easier than it had previously been for Greenpeace Nordic to avoid confronting the pathogenic organizational double bind and the confusing, painful emotions connected to it. The fact that Greenpeace Norway was now part of a firmly established, financially self-sustaining office with a strong voice in the international Greenpeace organization helped to alleviate the pain of the anti-whaling campaign. It also helped to convince international Greenpeacers to leave Greenpeace Nordic alone with the anti-whaling campaign in Norway and to stop demanding direct actions and so on every summer. It was hence not absolutely necessary for Nordic Greenpeacers to continue following an uncomfortable therapeutic prescription, solving the relational problem they had with their organization and developing their identities. This is John's judgment as stated in 2006:

Juliane: Why did they stop doing big actions on whales in 2000?
John: 'Cause they failed to understand what they were doing. They missed it. They took the coward's way out.

Juliane: They're saying, we've turned over the leaf of whaling, we can campaign on forests now in Norway, wow!
John: I think we did turn it over. And I think it's turned back. I think there was an understanding in the organization, and I think that disappeared. Has fluffed away in the fluffy fluffiness that the whale campaign is now. You can tell the truth a couple of times and people get it, but then if you don't carry on repeating that truth, they lose it again.

This illustrates the Luhmannian lesson that the *eigen*-dynamics of the social system, the dominance of communication structures, can be more powerful than an understanding inside individual psychic systems (see also Willke 1999). It is simplicity bordering on stupidity to portray individual people as manager heroes who solved an organizational problem (see Czarniawska 2005 for a critique of this). If a manager cannot get the social system to get it, to buy it, to understand the solution he or she puts forward, the solution will not work. Organizational problems must be solved by the organization. Social systems are not simply controllable and programmable by individual psychic systems.

The trouble with the current approach, if we believe Nils and John and some of my other interviewees, is that while it does allow Greenpeace to keep up its work in Norway, it will not enable Greenpeace to either ever win the whale campaign or ever win the support of Norwegians. That is the nature of a pathogenic double bind: you can go on forever without ever reaching your contradictory goals.

Nils: I believe Greenpeace Nordic have lost what they had won there, actually.

John: The organization's created this huge great big monster. Greenpeace is a horrible, beasty, evil thing in the minds of the Norwegians. There is no way that suddenly, this irrational monster is gonna come round and talk to people rationally.

The thing to do is to find someone else to be rational about the whaling issue. Like newspapers, like other organizations who're believed, like other people who're trusted.

And no one has said to the Norwegian campaigners, you're never going to win another campaign until you win the whale campaign. And *that* is the *biggest truth* that this organization needed to get hold of, that you *could not* campaign about *all* the big forest issues, all the other issues, until you win the whale campaign. When the Norwegians start to realize that they're not completely right about whaling, when the

myth that the Norwegians are the poor little hunted little poor little people, is blown open, all the other campaigns are gonna be completely open for winning. 'Cause when Norwegians start realizing that these animal-loving freaks actually had a point, and were right about something, they're gonna have to listen to the other things that we've been saying at the same time.
Juliane: Norwegian Greenpeacers believe they can campaign about other things now, although they haven't won…
John: They can't, it's like, Charles Manson campaigning for women's rights or something, you know –
Juliane: Who is Charles Manson?
John: (*laughs*) He's a mass murderer. You know, it's just the same sort of thing, it's like, you're completely insane, you're tree-huggers, you're the worst of lunatic fringe of environmentalism.

Greenpeace's Self-description, Again

The experience of the 1999 'big push' does suggest that a therapeutic organizational double bind can be successful and overcome a pathogenic organizational double bind. Nils puts it as follows: 'All those who were there back then must have seen that there was something that worked, at least. So I believe Greenpeace has learned something which is left'. Today, Nordic Greenpeacers feel less defined by, and angry and paranoid about, the whaling issue when working in Norway (witness Alex's judgment comparing the 'big push' to a person suffering from arachnophobia picking up spiders). They have a clearer idea of what they should be and do. They understand that Greenpeace cannot achieve its goals in Norway by merely putting pressure on Norwegians, and must aim to function in and with the Norwegian environment, presenting arguments that are apt to convince Norwegians. Greenpeace Nordic's relationship with the international Greenpeace organization seems to be more constructive. This may be a sign that Greenpeacers from different countries have greater emotional understanding of each other and of Norwegians today than they had before 1999.

However, the relational problem between Nordic Greenpeacers and their organization that has always been at the root of the trouble with the anti-whaling campaign has been mitigated, but not solved. Greenpeace in Norway has abandoned Nils and John's solution to this problem: convince Norwegians that whaling should be stopped and do not care if they dislike you. Greenpeace Nordic profits from double-binding itself, in that this enables it to maintain organizational self-descriptions that it is emotionally attached to: Greenpeacers are 'fixers of environmental problems', 'solvers of issues', efficient and effective doers. Those who work for Greenpeace in Norway, today, want Greenpeace to be heard. They want to present Greenpeace's position on issues (other issues than whaling). They want Greenpeace to achieve specific results, such as a ban on certain chemicals. They believe that a necessary condition for this is that Greenpeace is seen as knowledgeable, respectable and legitimate. They want to avoid coming across as emotional and irrational, as they believe they would if carrying out direct actions against Norwegian whaling.

Greenpeace Nordic's organizational self-description has changed since the old days, in that Greenpeace Nordic has understood that it is not very useful to be tough High North Frontliners all the time if the results of this are inadequate. But the organizational self-description has not changed at a fundamental level so as to reflect what John describes as central to his – and Nils' – philosophy: 'I think it's about changing the world, the way people feel about their place in society, and what they can do, what they can achieve'.

Current Greenpeacers do want to mobilize people, but the goal is not to empower people. It is to get politicians or other decision makers to decide in a certain way and not another because of the public interest and possibly the public pressure. In sum, Greenpeace Nordic does not 'work to become unnecessary' in Norway, as the old Greenpeace slogan would have it. Of course it wants to achieve results; but it also wants to maintain itself. Greenpeace Nordic would not today take a stance on a campaign which said that 'If you think Greenpeace is a bunch of idiots and hairy hippies, you think it. That's not the issue here. You don't have to like us, but we're still right'. Greenpeace Nordic's anxiety about coming across as emotional and irrational may also be an indication of an unwillingness to do difficult emotional work, address difficult questions, for example about organizational self-descriptions. I will come back to this in the conclusion.

John and Nils are not the only ones who would like Greenpeace to go back to being People Persuaders and 'work to become unnecessary'. The following quote from Alex expresses a longing for the '*ursprüngliche*' campaigning that changes people, not policies, the kind of campaigning that was represented by Nils' radical 'Ultimate Campaign Department' or '*Phyllis Cormack* Campaigning' vision.

> *Alex*: There's been changes in the organization globally also, very significant ones, in management as well as in structure. And that has of course been reflected also here.
>
> They don't talk about campaigns anymore. They talk about issues. Which I think is wrong. Whatever. I'm an old fart, I'm a real dinosaur,

> you know, I've been around here for too long, so maybe I'm just reacting negatively to new things, but there is a really different way of thinking about our work now compared to the way it was back in the mid-'90s...

It is of course not a priori implausible to argue that in order to achieve the results that would make Greenpeace unnecessary, Greenpeace first needs to be established as sensible and trustworthy. It is further not a priori implausible to argue that the current strategy has saved more whales in Norway than any other strategy, despite the fact that quotas have been increased. Ultimately, it is impossible to finally prove or disprove these arguments, as counterfactual evidence cannot be provided. This brings us to the Luhmannian conclusion that all our conclusions are necessarily contingent.

Conclusion

Fuck Greenpeace, but Save the World

It is possible that Nils and John are wrong. Maybe the pathogenic organizational double bind of Greenpeace in Norway – the organization cannot gain support because Norwegians dislike its opposition to whaling, but also cannot stop whaling because it does not have Norwegians' support – cannot ultimately be solved in the way that Nils and John envisioned it would be solved.

In 1999, after Greenpeace had launched its fresh criticism of Norway for breaking international agreements, the Norwegian Whaling Commissioner insisted that Norwegian whaling was entirely legal, because Norway had lodged a reservation against the IWC moratorium. 'That Greenpeace claims the opposite just shows how unsound their whale actions are' (Johansen 1999, my translation). Norwegians might ultimately have followed their government's argumentation and come to the conclusion that Greenpeace's argument that Norway was violating international agreements was wrong, meaning that there was no problem with Norwegian whaling. Thus, it is possible that Greenpeace would not have won the anti-whaling campaign by using Nils' and John's strategy. As John says, 'Maybe the strategy Nils and I developed hasn't worked. We never did the follow-up research we should have done, to find out how the actions that we'd done that year were perceived'.

It is also possible that Norway would have stopped whaling if Greenpeace had continued to use Nils' and John's strategy, because Norwegians would have accepted Greenpeace's argument about international agreements. But Norway might then have restarted whaling if the IWC had lifted its moratorium. In other words, Greenpeace might have won only a temporary victory by using Nils' and John's strategy.

The 'Fuck Greenpeace, but save the whales' strategy may thus not be the way to stop whaling in Norway. Greenpeace's current strategy has clearly not stopped whaling either. Consequently, the only way out of the Norwegian dilemma for Greenpeace may be to accept whaling as long as it is sustainable. This position would be in line with Greenpeace's character as an environmental, rather than an animal rights organization. However, it would in all probability cost Greenpeace support around the world. This is because many people who see whales as symbolic and special cannot accept the argument that it might be defendable, from an environmental point of view, to kill a limited number of (minke) whales.

If Norwegian whaling cannot be stopped and Greenpeace wants to avoid the negative consequences that accepting limited whaling would have, then Greenpeace Nordic's current strategy for campaigning in Norway could be interpreted to be the best option available. It may be the best way to live with a dilemma which cannot be escaped. Murphy and Dee (1992), analysing a conflict between the chemical company Du Pont and Greenpeace, come to the conclusion that it may have been functional for Greenpeace (and Du Pont) to not resolve the conflict in the end. Maybe the case of the anti-whaling campaign in Norway is similar.

It is of course also possible that, quite independently of whether or not Nils' and John's strategy was the right strategy back in the late 1990s, it would no longer be the right strategy today. Any Greenpeace campaigner who decided to have a new go at the whale campaign in Norway today would have to do some thorough research, a power analysis, etc. to determine what the situation is today.

In my opinion, the potential of the 'Fuck Greenpeace, but save the whales' strategy to solve Greenpeace's pathogenic organizational double bind in Norway was not realized, because the strategy was abandoned so soon. Let us recall that Nils and John believed that Norwegians could be persuaded that Norway should stop whaling, which would mean that Greenpeace would win the whale campaign and that consequently the campaign would no longer be a reason for Norwegians to dislike Greenpeace. The 'Fuck Greenpeace…' strategy may indeed have been the best strategy not only to achieve the campaign goal (persuade Norwegians to ban whaling), but also to win more popular support for Greenpeace in Norway. And it may have been the best strategy to win more support for Greenpeace, not although, but because it prioritized goal achievement over winning support for Greenpeace.

It is possible that protest organizations are the most paradoxical of all social systems, in that they are best at ensuring their own survival when they care least about it. Persuading people to support a cause and to act on an issue requires, firstly, an understanding of your opponents and their reasons, emotions and motives, before you attack them. It then requires taking the point of view of those people who may be persuaded to support your cause, trying to feel what they feel. It requires speaking to these potential supporters in a way that will resonate with them, that will activate those values they hold that will induce them to support your cause (see Lakoff 2004; Popovic and Miller 2015). In order to manage this, a protest organization needs to have local people on the ground in the relevant places and then to actually listen to what they say. It needs to take its employees' emotions seriously. It needs to take a reflective attitude, in the Luhmannian sense, to observe itself as a contingent unity in the environment it operates in. The less self-absorbed, the less occupied with maintaining its own self-descriptions and autopoiesis a protest organization is, the better it may achieve this, and the better it may be able to persuade people of its cause. And if people are persuaded to support a cause, they will often also support the organization representing that cause.

The 'Fuck Greenpeace, but save the whales' strategy was designed on the basis of a thorough understanding of the Norwegian opponents and audience. It resonated with the audience, activated those values Norwegians held that might induce them to oppose whaling. It put Norwegians in a double bind: you can either choose to go on whaling, preserving your self-description of small oppressed Askeladen-David, and be perceived by the rest of the world as a butchering Goliath who breaks international agreements. Or you can examine your emotions and question your self-descriptions, stop whaling and be seen by the rest of the world as a grown-up responsible member of the world community. Had Greenpeace kept Norwegians in this binding, there might well have been some 'working through' on the part of Norwegians, and Norwegians might not only have stopped whaling, but also have begun to support Greenpeace. If this is correct, then the fact that the 'Fuck Greenpeace…' strategy was abandoned so soon is regrettable both for the sake of the whales and for the sake of Greenpeace in Norway.

Greenpeace Nordic today prefers a pathogenic organizational double bind situation to solving the whale campaign problem in Norway once and for all, one way or another. It hopes the problem will solve itself as whaling in Norway dies a 'natural death'. I interpret this as unwillingness on the part of Nordic Greenpeacers to examine their own emotions, to question their organizational self-descriptions. Nils and John are no longer there to force them into such an examination, or to offer them emotional compensation for any losses they might experience in the process.

John doubts that his and Nils' work in 1998–1999 ultimately had any truly therapeutic effect on Greenpeace at all. He thinks that Greenpeace may have gone along with the 'Fuck Greenpeace…' strategy in 1998–1999, not in order to actually solve the whale campaign problem properly, but out of a desire for some kind of fix for the problem. He thinks that Greenpeace's organizational behaviour after he and Nils left in 2000 should be interpreted as a 'total relapse' into the pathogenic organizational double bind, into old organizational behavioural patterns.

The case of Greenpeace in Norway thus demonstrates that it is difficult for protest organizations to develop a reflective attitude. Most systems, such as organizations, will experience increased awareness of their own contingency as increased insecurity and as uncomfortable. There is reason to assume that reflection is particularly difficult for not-for-profit organizations working towards idealist goals, compared to other types of organization. It takes strong motivation and conviction and a lot of energy to do the work they do, to develop and maintain self-descriptions that contradict society's dominant self-descriptions and to work towards ideals that are not dominant in society. So they have less space and energy for questioning their own motivations and actions and for understanding others' motivations and convictions. But, also, for an organization that is trying to save the world, reflection means remembering that it must strive to become unnecessary. A protest organization striving for reflexivity must, over and above coping with the normal degree of insecurity and discomfort associated with organizational reflection, search for ways to (in the long run) make its own autopoiesis redundant. This is completely at odds with what social systems usually do, which is to strive to continue self-reproduction. It is an entirely un-Luhmannian, and probably scary, thing for a social system to do. (On the tension for not-for-profit organizations working towards idealist goals between the fact that they exist solely to achieve something which is more important than their own existence and the inherent tendency that they share with other organizations to prioritize maintenance of their own autopoiesis over other goals, see also Dale 1996: 60ff.)

Did Greenpeace Nordic, and Greenpeace as a whole, find reflection on the whale campaign in the late 1990s difficult because it found it difficult to envision a world where Greenpeace is not needed anymore to campaign against whaling? The following quote by Nils in 2005 points in this direction.

> *Nils:* I believe there are two completely different levels there. One level is like – and unfortunately I believe that many Greenpeacers stay on that level (*chuckles*) –, as long as Greenpeace works with the whale campaign everything is fine, you know. But the sick thing about that is that, then you're not allowed to win the campaign, because then you wouldn't have a whale campaign anymore. I mean there is no-one who thinks like that, consciously, but that's a bit what's showing in the patterns of the campaign work that's being done.

Of course, a protest organization which actually does achieve its goal (such as stopping whaling) may continue its autopoiesis by working towards a new goal. If a protest organization strives for reflexivity, this does not mean that it will inevitably, immediately or eventually, make its own autopoiesis redundant. But for a protest organization, reflection will heighten awareness of a possible future where the organization no longer exists, at least not in its current form. This may evoke intense emotions (in addition to the strong emotions that often exist around such organizations anyway), because emotions are signals which indicate if social systems are threatened.

While the example of the Greenpeace whale campaign in Norway shows that it is difficult for protest organizations to develop a reflective attitude in the Luhmannian sense, it also suggests that it is important for them to do so if they want to reach their stated goals. Greenpeace was so busy with its own self-descriptions that came with troublesome blind spots, with its internal campaigning against each other, etc., in short, so busy with itself all the time that its effectiveness in Norway was appreciably reduced. If social movement organizations develop a reflective attitude, they will be more able to recognize their own blind spots and to induce reflection and change in society. That of course is the whole point of their existence.

It is possible that if the international organization Greenpeace developed a Luhmannian reflective attitude, if it reflected about how it can become unnecessary, it would come to the conclusion that it should not campaign against whaling anymore. Let us recall that, in the early days of Greenpeace, Bob Hunter and other representatives of the faction of 'mystics' (Zelko 2013: 211) within Greenpeace saw the whale campaign as an important part of their campaigning to bring about a new, ecological mass consciousness. The older, sober Greenpeace activists opposed the idea of the young organization diverting energy from the anti-nuclear cause to the 'soft' issue of whale protection (Zelko 2013: 174ff.; see also Weyler 2004: 230; Hunter 2011). It is possible that whales have indeed served as a 'gateway species' which made many people around the world aware of how precious and how vulnerable nature is, how necessary it is that we stop destroying our natural environment and start living in harmony with it. But it is also possible that the campaigning for the protection of whales that has been done over the past decades has done more harm than good. It may have diverted attention away from more important environmental and political issues and the connections between these, from broader and deeper historical, political, structural analyses. It may have contributed to making environmentalism look simple and esoteric (see also Dale 1996: Ch. 10–11).

The nature of Greenpeace's campaigning, with its strong reliance on direct actions and making the headlines, may have contributed to this effect

(Dale 1996: Ch. 10–11). What is needed to achieve real change is long-term activism that truly educates the public (Dale 1996; see also for example Zinn 2009). The media stunts that Greenpeace is so good at can create space in which educating the public, working to encourage people to become critical citizens, is possible. But they are not in and of themselves sufficient. Organizational reflection might lead Greenpeace towards reducing the number of media stunts. It might also lead Greenpeace towards reducing its presence in Norway.

> *Juliane*: … Why not shut down Greenpeace Norway, start something in Africa.
> *John*: Oh why not shut down a whole load of Greenpeace offices and start something in Africa, I don't know. If shutting down Greenpeace Norway meant that we would have saved money for Africa, then I would have said let's do Africa instead. I know the international Greenpeace organization wouldn't have given me the option to do Greenpeace work in Africa instead of Norway. I think Greenpeace's work in the Nordic countries should be scaled right down (*laughs*) and the rest of the money should be spent in the places that really need it.
>
> *Alex:* Honestly, on a global stage, what are the Nordic countries compared to China? What are we compared to, like, in ten years, Africa?

We can never know what reality 'is really like'. Your point of view may differ from mine. The analysis presented in this book – the story I have told of the Greenpeace whale campaign in Norway – is contingent. You are welcome to present a different one if you want to. I hope that by reading this book you have gained a deeper understanding of the Greenpeace whale campaign in Norway than you had before. In particular, I hope that by reading this book, you have gained a deeper understanding of the organization-internal dynamics behind this campaign than you had before. I hope this book will contribute to a continuing effort to understand how social movement organizations succeed in achieving social change, or fail to do so. I also hope to have shown that systems theory is useful for the study of protest.

> *Alex*: The universe is unimaginably huge. And in this whole giant universe there's, as far as we know, only one place which has life. And it's this, like, little thin layer (*indicates with hands*) around a ball of dead rock. And it's surrounded by hostile environment everywhere else, as far

as we know that's the only place where there's this incredible rich multitude of species and life forms and it's so small, as far as we know.

And what kind of hubris, what kind of balls does it take to say, well, we can destroy some of it. Doesn't matter. Because this way, it's gonna, you know, increase shareholder value. I mean, that is just like, (*laughs*) not on, you know.

That is my system of beliefs. Which I think Greenpeace espouses, which are not espoused by the economics or the prevalent philosophy. So, there are these general values that need to be translated into all the different cultures in the world.

And I think it's possible to do that.

Appendix

Some Additional Systems-theoretic Explanations

As has been explained in the introduction, we should not think of cognition as an input-output model (as in, the environment gives an open system information). We should use a concept of operationally closed systems (Maturana and Varela 1987: Ch. 6, 176ff.; Bakken 2000: 74f.; Simon 2000a: 79). Maturana and Varela used the term autopoiesis (Greek: autos = self, poiein = to make) to describe living systems characterized by the ability to produce and reproduce the elements they consist of. An autopoietic system is operationally closed because the operations leading to the productions of new elements in the system are dependent on earlier operations of the system and are the basis for the following operations (Maturana and Varela 1987: 50ff.; Maturana 1999: 153f.; Baraldi, Corsi and Esposito 1997: 29f.; Luhmann 1997b: 65ff.).

Luhmann generalizes the concept of autopoiesis and applies it to psychic and most importantly, social systems (Luhmann 1997b: 66; Baraldi, Corsi and Esposito 1997: 29f.). Thus, the elements of psychic systems[1] are experiences (instances of *Erleben*). Experiences can be conscious or unconscious.[2] The psychic system produces an experience, then produces a new experience and connects it to the previous one, then produces and connects the next experience, and so forth. It reproduces its elements all the time. If it stops doing so, it no longer exists. The psychic system cannot transfer its experiences to another psychic system, or to a biological or a social system. There are no experiences outside of psychic systems (Luhmann 1984: 354ff.; Baraldi, Corsi and Esposito 1997: 29f.).

The elements of social systems, in turn, are communications.[3] A social system produces a communication, then produces a new communication and connects it to the previous one, then produces and connects the next communication, and so forth. It reproduces its elements all the time. If it stops doing so, it no longer exists. The social system cannot transfer its communications to a psychic system, because there are no communications outside of social systems. Psychic systems are unable to communicate (they only experience). Psychic systems are therefore part of the environment of social systems, just like, say, trees (Baraldi, Corsi and Esposito 1997: 29f.;

Luhmann 1984: Ch. 4). However, psychic systems are necessary for the emergence of social systems.

As has been explained in the introduction, information in social and psychic systems is a system-internal quality. Because an autopoietic system is operationally closed, information cannot be in the system's environment and then be transferred into the system as an 'input' to which the system is open. However, while psychic and social systems are operationally closed, they are cognitively open.[4] External stimuli can serve as perturbations to an autopoietic system, which are then processed by the cognitive structure of the system.

Psychic and social systems can and do influence each other, in the sense of mutual perturbation. In fact, because they have developed in coevolution, they are particularly well suited to perturbing each other. Both social and psychic systems are systems of meaning (*Sinn* in German); they draw their boundaries and connect their elements on the basis of meaning (Luhmann 1984: 92ff.; Luhmann 1997b: 44–59 and 110f.; see also Maturana 1999).

At the same time, social systems possess autonomy; they possess *eigen*-dynamics which cannot be determined by psychic systems. (Neither can social systems control or determine psychic systems.) In the language of organization studies, this implies that planned or managed change in organizations is difficult to achieve. This is exactly what we see in the case of Greenpeace in Norway. Nils and John knew that although they were leaders in the organization, they could not simply order the organization to do things differently in Norway. The organization would have to understand what Nils and John were saying and change its own dynamics; it would have to buy into Nils' and John's new strategy. Nils and John did not stay with the organization long enough, and keep up the therapeutic organizational double bind long enough, for this to happen properly.

Evolution, or unplanned change, in organizations and other social systems happens all the time, however. This is because the communications social systems consist of don't last. They vanish the moment they have happened. Thus, social systems can only continue to exist if every one of their communications is followed by a new communication (Luhmann 1984: Chapter 4; Luhmann 1997b: 81ff.; Luhmann 2006: 53; Baraldi, Corsi and Esposito 1997: 91). Because new communications need to be produced all the time, there are opportunities for alterations all the time. Social systems therefore can and do change all the time. They are 'endogenously turbulent' (Luhmann 1990: 36, my translation).

Because social systems cannot receive direct informational input from their environment, they can never know what their environment 'is really like' (Luhmann 1995b: 34f.). Instead, in the process of (system-internal) cognition, the social system, on the basis of the external perturbations it processes, produces an internal under-complex reconstruction of what is outside of it (Luhmann 1984: 47ff.; Luhmann 2006: 314; Luhmann 1997b: 124).[5]

When doing empirical research in the social sciences using a constructivist lens such as Luhmann's theory, and observing systems, what we should strive for is *Verstehen* in the following sense: 'Understanding requires … that one interpret the system to be understood as a system that is meaningfully oriented to its own environment' (Luhmann 1995b: 88). The observer must be aware that the observed system is itself an operationally closed system producing its own meaning while drawing on perturbations from the environment. He or she must try to get an idea of what that meaning might be. This is 'ambitious', as Luhmann himself says (Luhmann 1984: 110f.). In the process of my research I tried to develop this kind of *Verstehen* of Greenpeace in Norway, of the organization's behaviour, problems, strategies and reflection.

For an autopoietic system, it is decisive just which differences make a difference for the system, what is information for the system (and what is merely white noise). What if the system misses crucial points about its environment? What if its internal reconstruction of its environment is under-complex in a bad way, such that it does not include things that are important for the continued reproduction – in other words, existence – of the system? (Luhmann 1984: 47ff.). What makes such situations particularly dangerous and difficult is the fact that a social system can miss a point *and* fail to understand that it has missed a point. In other words, a social system can miss environmental perturbations that indicate it is not picking up important environmental perturbations. (The analogous argument applies to psychic systems.)

Indeed, one of the main arguments in Luhmann's (1990) book *Ökologische Kommunikation* (Ecological Communication) is that modern society has problems including in its communication the threats of ecological destruction in its environment. And one of the main arguments of *Protest* (Luhmann 1997c) is that the function of social movements, such as the environmental movement, in modern society is to bring such issues into communication. If the economic system or the political system fail to communicate about things which threaten modern society, social movements can alarm them, perturb them so as to include these things in their communications. Protest movements thus serve an immune system function for society; Luhmann calls their contribution 'fruitful' (1997a: 104, my translation). With luck, protest movements can bring about reflection in society: they can induce social systems to try and understand which important environmental perturbations they have missed, to try and examine their own 'blind spots'.

According to Luhmann, when protest movements produce their alarming communications, they assume the problems should be solved somewhere else (Luhmann 1997c: 188). They feel free not to take into consideration the internal workings of the social systems they are criticizing. They don't try to understand their opponents. They just use the form of protest and try to make the opponents accept the protest view with the help of drastic means such as direct actions: Do as we say, otherwise X (something bad) will happen (Luhmann 1997c: 205f.; 178). Compare this to Nils' statement, quoted in Chapter 4:

> *Nils*: It's not just Greenpeace, it's very common in a campaign group that you make your plans based on your own world view. It is very unusual that campaign groups go out and try to understand, what do you think and how do you view this. What are you interested in and what can I do to make you interested. A traditional campaigner paradigm, I believe, if you can talk about such a thing at all, is like, you know roughly what you want and then you take the tools you usually use. And then you drive hard, like, adamant.

This is exactly what Greenpeace, an autopoietic, operationally closed social system which is part of a social movement, did in Norway. This led to the intriguing situation that Greenpeace, an autopoietic social system which intended to induce other autopoietic systems to examine their blind spots, failed to notice important environmental perturbations, and failed to notice that it had failed to notice important environmental perturbations.

The fact that this resulted in a notable lack of success for Greenpeace in Norway for a long time did not really make the organization reconsider its strategy (at least not in a comprehensive fashion). It is often taken for granted that if a social system does not reach its goals, it will reflect on the reasons why and change its strategy. Not so, says Luhmann: purpose is not a sufficient guide for action or reflection. Building on March and Simon (1958: 165), Luhmann says that purposes, such as stopping whaling, are not motives which allow one to understand and explain organizations' operations, although organizations may use a purpose as a justification for certain decision chains.[6] The main 'purpose' or 'goal', the first priority, of an autopoietic system such as an organization is to continue operating, to maintain its own autopoiesis (Luhmann 2006: 165, 183–185, 256f.). What Greenpeace did in Norway did serve the function of enabling Greenpeace's own autopoiesis. As long as Norway did not stop whaling and the 'David against Goliath' and 'Campaigning against each other' loops sustained themselves, Greenpeace's self-reproduction (with regard to the whale campaign) was secured. So the fact that it did not reach its goal did not suffice to make the social system Greenpeace change. And for a long time, it was not obvious to Greenpeace – because of the organization's own blind spots – that what Greenpeace did in Norway was not the right thing to do to reach the organization's goals.

According to Luhmann, reflection happens when, instead of unquestioningly continuing to reproduce itself, the system observes itself as a contingent unity in an environment (Luhmann 1995b: Ch. 11). When observing itself as a contingent unity in an environment, the system is, potentially, able to compare this unity with alternatives. It is, potentially, able to realize whether it has failed to pick up on important environmental perturbations, and to choose to operate differently in the future so as to take these into account.

Rationality, according to Luhmann, is reflection under the aspect of the unity of system and environment (Luhmann 1984: 617). The issue here is that the system has to survive in its environment: 'Translated into the language of causality, … a system must control its effects on the environment by checking their repercussions upon itself if it wants to behave rationally. A system that controls its environment in the end controls itself' (Luhmann 1995b: 475).[7]

Achieving reflection and rationality is a paradoxical task because a system has to reflect – observe itself as a contingent unity in an environment – using its own information processing structure. All the system can know about the environment is the under-complex internal reconstruction of the environment it has produced itself; the system cannot step outside of itself to see 'what the environment is really like'. And the system cannot step outside of itself to observe itself, either. It can only know its own cognitive structure through using this very same imperfect and blind-spot-riddled cognitive structure. For this reason, ultimately the only thing the system can be sure of about itself is that it is 'what the environment is not'. But this means that for the system, the environment is what the system constructs it to be, and the system is what the environment is not. How can the system observe itself as a contingent unity in its environment if this is the best it can know about itself and the environment? How is the system supposed to control its effects on the environment by checking their repercussions upon itself, in other words, achieve 'fit' with its environment in order to survive in its environment, if it does not know what either its environment or itself 'are really like'?

Vos (2005) argues that this problem can only be solved through deparadoxization (see Figure A.1). Outside-in self-observations assume they know or realize something about the environment which is true, and then observe the system (for example the organization) in relation to that. Inside-out self-observations assume they know something about the system which is true, and then observe the environment in relation to that. The two approaches are equally valuable solutions to the same problem: that there are no privileged points of observation, no positions where the system can stand and be sure that it sees and understands 'true reality' (Vos 2005; cf. also Willke 1999: 172 and Czarniawska 2005).

Strictly Luhmannianly speaking, a system which uses an outside-in observation strategy or an inside-out observation strategy in order to reflect and to achieve rationality is fooling itself, because it cannot know whether that what it assumes to be true about the environment or about itself really is true. But 'radical constructivism', where the system is entirely aware of the contingency of all its operations, is in practice unachievable. It would lead to collapse. At some point, the system must assume that something is 'true' in order to be able to function, reproduce itself and reflect (Luhmann 2006; Vos 2005).

Greenpeace used both an inside-out and an outside-in self-observation strategy to reflect on the anti-whaling campaign in Norway. The 'scuffle' between Nils and John that has been described in Chapter 4 was an inside-

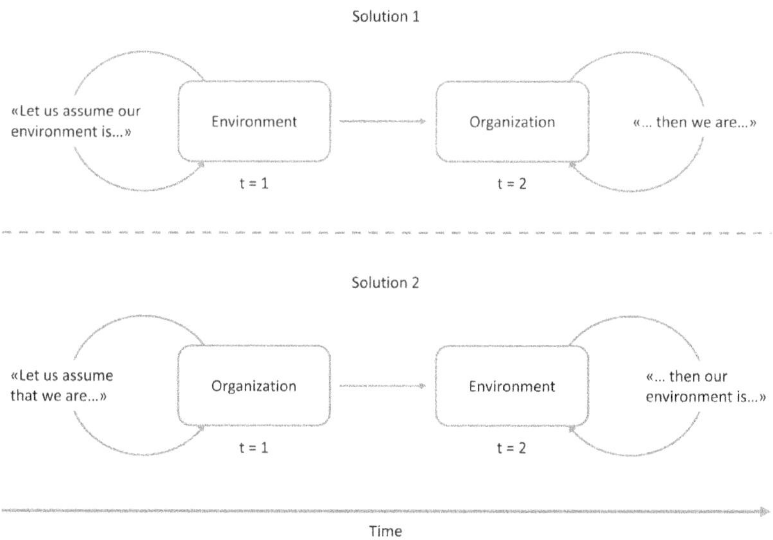

Figure A.1. Solving the problem of self-observation by using deparadoxizing strategies
Source: Vos 2005: 373, Figure 3

out self-observation strategy: 'Let us assume that we are… then our environment is…'. Who are we and why are we doing this? Why do we stand for this? What is our self-description that makes us do it?

The self-description of the Greenpeace system was defined as People Persuaders with a rational, environmental reason for wanting to stop whaling. It was then possible to look at what the environment to this kind of organization was ('the environment is what the system is not'): one which the organization must persuade that there are rational arguments against whaling. The self-description has formulated the problem in such a way that it begs a communication strategy explaining principles that are generally understandable to a majority of people. Proceeding from the self-description they had agreed on, John and Nils could deduce how the environment should be treated, which is what an inside-out deparadoxizing strategy is about (see Figure A.2).

The focus group study John ordered in 1998 was an outside-in deparadoxizing strategy: 'Let us find out what the environment is; that will tell us what we have to be'. Nils and John could not see 'reality as it really was'. But they could ask Norwegians what their reality was. They were aiming for *Verstehen* in the ambitious Luhmann sense. They wanted to understand the views, emotions and self-descriptions of Norwegian society.

The outside-in deparadoxizing strategy was successful, in that it made it possible to derive clear instructions along the lines of 'If our environment is …, then we are …' for the whale campaign. If the Norwegian environment thinks international agreements should be respected, then Greenpeace should highlight what it regards as Norway's breach of international agreements about whaling. On the other hand, if Norwegians think Greenpeace is strange and irrational, then Greenpeace is not the rational voice that can tell them to stop whaling.

With their inside-out and outside-in deparadoxizing strategies, Nils and John were striving for reflexivity, in the systems-theoretic sense. Instead of

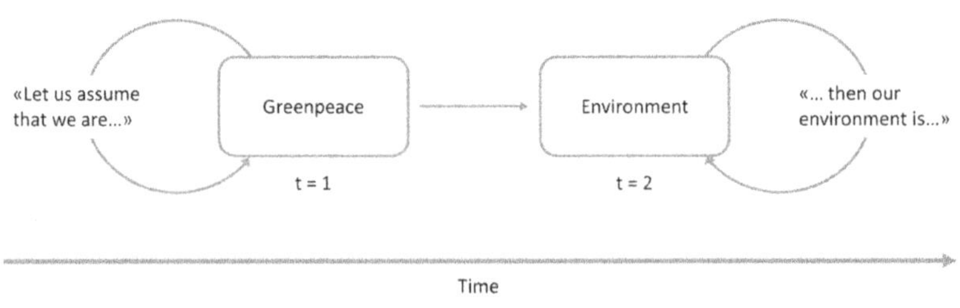

Figure A.2. Nils' and John's inside-out deparadoxizing strategy
Based on: Vos 2005: 373, Figure 3

just taking the existing Greenpeace-internal reality constructions at face value and assuming that this was reality, Nils and John tried to find out whether the constructions could be corrected, improved.

And ultimately, Nils and John were striving for rationality, in the systems-theoretic sense. Nils and John wanted the social system Greenpeace to function properly in the Norwegian environment, to be supported by the Norwegian environment. So they were trying to achieve a 'fit' with the Norwegian environment. They were trying to find out what the effects of Greenpeace's actions on the environment were and how Greenpeace would have to treat the environment in order to achieve those repercussions of its own actions that it sought to achieve.

John and Nils tried to motivate the social system Greenpeace, through their therapeutic organizational double bind, to go through the same process of reflection they had gone through. They presented the results of the focus group study to their colleagues, asked them to talk to Norwegians in Oslo's main station, etc. They tried to enable Greenpeace Nordic to critically examine its organizational self-descriptions and the emotions attached to these as they themselves had done. As we know, they did not quite succeed in this. Yet the temporary and incomplete results they achieved indicate that social systems such as Greenpeace can in fact reflect.

Luhmann discourages belief in the control and manageability of social systems. Yet he emphasizes the importance of self-observation of social systems that is done in a more (systems-theoretically) informed way, with more emphasis on context and contingency than management theories generally offer. A social system can become self-critical, aware that what it takes to be reality is an internal construction. It can become humble in the sense of giving up the belief that one could ever reach the 'right position'. Using the deparadoxizing strategies described above, a system can observe itself as a contingent unity in the environment. Then, it can attempt to judge whether the unity is normal or pathologic. It can try to see whether its own

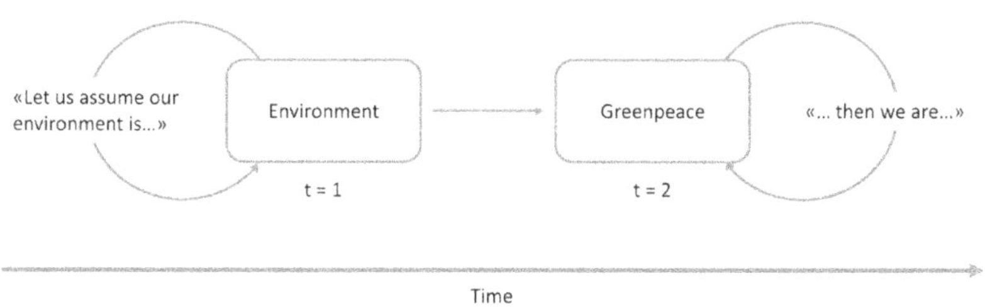

Figure A.3. Nils' and John's outside-in deparadoxizing strategy
Based on: Vos 2005: 373, Figure 3

structures, which are riddled with blind spots, lead to a distorted perception of 'reality' – maybe a 'bad' treatment of the environment (Luhmann 2006). The less preoccupied with maintaining its autopoiesis and self-descriptions the system is, the better it will be at this paradoxical task.

As I have explained in this book, it is a difficult task. It may be particularly difficult for protest organizations. One reason for this may be that for a protest organization, trying to achieve rationality will heighten awareness of a possible future where the organization no longer exists, at least not in its current form. Behaving rationally means functioning in the environment, being supported by the environment – but ultimately, if the environment supports what the organization stands for, this means the organization becomes unnecessary.

However, while reflection may be difficult for social movement organizations, if they want to be effective they need to reflect from time to time. Social movements function as an immune system for society. They alarm and perturb, they force society to put neglected problems on the agenda and hopefully to keep these problems on the agenda until they are solved. They make society reflect and change. If they do not achieve this, because they do not understand their audience, because they engage in counterproductive activities, etc., then what is the point of social movements?

Reflection and change in society take many different forms. It is difficult to judge a social movement's outcomes, effects and consequences. If people who are involved in a social movement develop a social identity, a collective self-description which is different from and hopefully more viable than society's dominant self-descriptions, this already constitutes a change in society. This kind of change is not easily achieved, because it is difficult to think and live differently. Social movements, and social movement organizations, are places where people learn, communicate, change their own minds and lives, etc. Thus, social movement organizations can effect change in society without achieving their stated goals. But in order to achieve these goals, they need to prioritize the finding of ways to persuade others over their own autopoiesis and reproduction of their own self-descriptions.

Notes

1. The term 'psychic system' is sometimes used interchangeably in the systems-theoretic literature with the term 'system of consciousness'. I choose the term 'psychic system' because it can comprehend processes which are not, or not completely, conscious.
2. Luhmann speaks of the elements of psychic systems as thoughts. Following Wasser (2003), I change this assumption to state that the elements (or basic operations) of psychic systems are experiences, because this can include unconscious as well as

conscious operations. Wasser indicates that Luhmann did not object to conceptualizing experiences instead of thoughts as the elements of psychic systems. Cf. Luhmann: 'However one wants to define the elemental units of consciousness (we will leave aside the distinction between ideas and sensations and speak of thoughts)' (Luhmann 1995b: 262; in the German edition Luhmann 1984: 355f.). My choice here is obviously related to my preference of 'psychic system' over 'system of consciousness'.

3. For details of Luhmann's concept of communication, see Luhmann 1984: 193ff. and the very good and intuitive explanation in Baraldi, Corsi and Esposito 1997: 89ff. The elements of organizations are decisions. Organizations are a special type of social systems, and decisions are a special type of communications (Luhmann 2006).
4. This is a departure from Maturana's very 'closed' conceptualization of autopoiesis; see Bakken 2000: 147.
5. Luhmann (2006: 77f.) points out that Meyer and Rowan (1977), who use a different theoretical framework, make statements with rather similar implications.
6. Luhmann (2006: 165) quotes March's idea that organizations are systems that search for purposes (see March and Olsen 1976).
7. Luhmann refers to Wilden (1972) here: 'the system which disposes of its environment disposes of itself' (p. 207).

Bibliography

Aasjord, B. 1993. 'Hvaldebatten og Naturvernforbundets rolle', in N.C. Stenseth, A.H. Hoel and I.B. Lid (eds), *Vågehvalen – valgets kval.* Oslo: Ad Notam Gyldendal, pp. 167–172.

Aass, H.P., and H. Bergsaker. 1994. 'Filmbevis mot hvalaksjonister', *VG*, 25 July, 9.

Abeles, G. 1976. 'Researching the Unresearchable: Experimentation on the Double Bind', in C.E. Sluzki and D.C. Ransom (eds), *Double Bind: The Foundation of the Communicational Approach to the Family*. New York, London and San Francisco: Grune & Stratton, pp. 113–149.

Ahlemeyer, H.W. 1995. *Soziale Bewegungen als Kommunikationssystem. Einheit, Umweltverhältnis und Funktion eines sozialen Phänomens.* Opladen: Leske & Budrich.

Aminzade, R., and D. McAdam. 2002. 'Emotions and Contentious Politics', *Mobilization* 7(2): 107–109.

Andenæs, U. 1987. 'Felles kamp for miljøet', *Aftenposten Morgen*, 28 April, 4.

Andersen, A.G. 1983. 'Alle tiders rekord for fiskerinæringen: Eksporterte for 7,5 milliarder', *Aftenposten Morgen*, 23 December, 5.

Andersen, B., and E.H. Indsetviken. 2014. 'Jakten på vågehvalen har startet', *NRK*, 26 April. Retrieved on 24 August 2014 from http://www.nrk.no/nordland/arets-hvalfangst-i-gang-1.11684943.

Andersson, A. 2000. 'Brevaksjon mot hvalfangst', *Bergens Tidende Morgen*, 26 April, 8.

Andresen, S., and T. Skodvin. 2008. 'Non-state Influence in the International Whaling Commission, 1970 to 2006', in M.M. Betsill and E. Corell (eds), *NGO Diplomacy. The Influence of Nongovernmental Organizations in International Environmental Negotiations.* Cambridge, MA and London: MIT Press, pp. 119–147.

Argyris, C. 1977. 'Double Loop Learning in Organizations', *Harvard Business Review* September-October: 115–125.

———. 2006. *Reasons and Rationalizations: The Limits to Organizational Knowledge.* Oxford: Oxford University Press.

Aron, L. 1991. 'Working Through the Past – Working Toward the Future', *Contemporary Psychoanalysis* 27(1): 81–109.

Ashforth, B.E., and F. Mael. 1989. 'Social Identity Theory and the Organization', *Academy of Management Review* 14(1): 20–39.

Baecker, D. 2004. 'Einleitung: Wozu Gefühle?', *Soziale Systeme. Zeitschrift für soziologische Theorie* 10(1): 5–20.

Bailey, J.L. 2009. 'Norway, the United States, and Commercial Whaling: Political Culture and Social Movement Framing', *The Journal of Environment & Development* 18(1): 79–102.

Bakken, T. 2000. *System og iakttagelse. Teorien om autopoietiske systemer og dens implikasjoner for moral og etikk.* Oslo: Det historisk-filosofiske fakultet, Universitetet i Oslo.

———, and T. Hernes (eds). 2003a. *Autopoietic Organization Theory. Drawing on Luhmann's Social Systems Perspective.* Copenhagen: Abstrakt Liber Copenhagen Business School Press.

———, and T. Hernes. 2003b. 'The Macro-Micro Problem in Organization Theory: Luhmann's Autopoiesis As a Way of Handling Recursivity', in T. Bakken and T. Hernes (eds), *Autopoietic Organization Theory. Drawing on Luhmann's Social Systems Perspective.* Copenhagen: Abstrakt Liber Copenhagen Business School Press, pp. 53–74.

Banaszak, L.A. 1996. *Why Movements Succeed or Fail: Opportunity, Culture, and the Struggle for Woman Suffrage.* Princeton, NJ: Princeton University Press.

Bandler, R., and J. Grinder. 1975. *The Structure of Magic I. A Book about Language and Therapy.* Palo Alto: Science and Behavior Books.

Baraldi, C., G. Corsi and E. Esposito. 1997. *GLU: Glossar zu Niklas Luhmanns Theorie sozialer Systeme.* Frankfurt am Main: Suhrkamp.

Barnes, J.A. 1954. 'Class and Committees in a Norwegian Island Parish', *Human Relations* 7(1): 39–58.

Barrett, G. 1993. 'Whalers Fight to Preserve a Way of Life', *The Age*, 25 May, 15. Retrieved on 11 March 2010 from LexisNexis database.

Bateson, G. 1983a. 'Double Bind, 1969', in G. Bateson (ed.), *Ökologie des Geistes. Anthropologische, psychologische, biologische und epistemologische Perspektiven*, 2. Aufl. Frankfurt/Main: Suhrkamp, pp. 353–360.

———. 1983b. 'Form, Substanz und Differenz', in G. Bateson (ed.), *Ökologie des Geistes. Anthropologische, psychologische, biologische und epistemologische Perspektiven*, 2. Aufl. Frankfurt/Main: Suhrkamp, pp. 576–597.

———. 1983c. *Ökologie des Geistes: Anthropologische, psychologische, biologische und epistemologische Perspektiven*, 2. Aufl. Frankfurt/Main: Suhrkamp.

———, D.D. Jackson, J. Haley and J.H. Weakland. 1956. 'Toward a Theory of Schizophrenia', *Behavioral Science* 1(4): 251–264.

———, D.D. Jackson, J. Haley and J.H. Weakland. 1963. 'A Note on the Double Bind – 1962', *Family Process* 2(1): 154–161.

Becker, K.H., and D. Seidl. 2007. 'Different Kinds of Openings of Luhmann's Systems Theory: A Reply to la Cour et al', *Organization* 14(6): 939–944.

Benford, R.D. 1997. 'An Insider's Critique of the Social Movement Framing Perspective', *Sociological Inquiry* 67(4): 409–430.

Bentzrød, S.B. 1992a. 'En "grønn Goliat" mot Norge', *Aftenposten Morgen*, 19 July, 4.

———. 1992b. 'Duellen', *Aftenposten Morgen*, 2 August, 2.

———. 1993. 'Greenpeace får hard kritikk i ny bok', *Aftenposten Morgen*, 5 June, 2.

Berntsen, B. 2011. *Grønne linjer: Natur- og miljøvernets historie i Norge*. Oslo: Unipub.

Beutler, L.E., C. Moleiro and H. Talebi. 2002. 'Resistance in Psychotherapy: What Conclusions are Supported by Research', *Psychotherapy in Practice* 58(2): 207–217.

Bigelow, A. 1959. *The Voyage of the Golden Rule: An Experiment with Truth*. Garden City, New York: Doubleday & Co., Inc.

Bjørnbakk, J.-M. 1997. 'Brann i hvalfangstskute påsatt', *NTBtekst*, 30 April.

Blok, A. 2008. 'Contesting Global Norms: Politics of Identity in Japanese Pro-Whaling Countermobilization', *Global Environmental Politics* 8(2): 39–66.

Bohlen, J. 2001. *Making Waves: The Origins and Future of Greenpeace*. Montreal, New York and London: Black Rose Books.

Borrevik, L.N. 1993. 'Gro må velge; hval eller EF', *Dagens Næringsliv Morgen*, 8 May, 10.

Bourdieu, P. 1990. *The Logic of Practice*. Stanford, CA: Stanford University Press.

———. 2001. *Masculine Domination*. Stanford, CA: Stanford University Press.

Bragli Alstadheim, K. 1992. 'Berthiniussen tar ikke avstand fra Sea Shepherd', *Klassekampen*, 25 July.

Brenner, C. 1987. 'Working Through: 1914–1984', *Psychoanalytic Quarterly* LVI: 88–108.

Broch, H.B. 2003. 'The Battle of Ålo, 1996. A Descriptive Analysis of a Confrontation Between Greenpeace Activists and Norwegian Minke-whalers', *Folk* 45: 177–206.

Brown, P. 1993. 'Norway Set to Defy World Whaling Ban', *The Guardian*, 19 May, 8. Retrieved on 11 March 2010 from LexisNexis database.

Brown, J. 1996. 'Greenpeace Targets Dangers of Fish "Hoovering"', *Press Association*, 29 April.

Brown, A.D., and K. Starkey. 2000. 'Organizational Identity and Learning: A Psychodynamic Perspective', *Academy of Management Review* 25(1): 102–120.

Brøyn, M.B. 1994. 'Greenpeace om hvalfangst: - Fryktet dette', *Aftenposten Morgen*, 11 September, 4.

Burnett, D.G. 2012. *The Sounding of the Whale: Science and Cetaceans in the Twentieth Century*. Chicago and London: The University of Chicago Press.

Burr, V. 2003. *Social Constructionism*, 2nd edn. London and New York: Routledge.

Bø, T. 1994a. 'Tre fra Greenpeace: Siktes for skadeverk', *Aftenposten Morgen*, 12 July, 4.

———. 1994b. 'Verbal hvalduell på "Solo"', *Aftenposten Morgen*, 26 July, 16.

———. 1994c. 'Test for vern av norsk hvalfangst', *Aftenposten Morgen*, 7 September, 12.

Carmin, J., and D.B. Balser. 2002. 'Selecting Repertoires of Action in Environmental Movement Organizations: An Interpretive Approach', *Organization & Environment* 15(4): 365–388.

Carr, A. 2001. 'Understanding Emotion and Emotionality in a Process of Change', *Journal of Organizational Change Management* 14(5): 421–434.

Cheek, F.E., and M. Di Stefano Miller. 1983. 'The Experience of Stress for Correction Officers: A Double-Bind Theory of Correctional Stress', *Journal of Criminal Justice* 11(2): 105–120.

Christensen, P. 2014a. 'Silda: fra kanonfiske til katastrofe', in N. Kolle (ed.), *Havet, fisken og oljen 1970–2014*. Bergen: Fagbokforlaget, pp. 15–31.

———. 2014b. 'Fra overflod til overkapasitet', in N. Kolle (ed.), *Havet, fisken og oljen 1970–2014*. Bergen: Fagbokforlaget, pp. 99–133.

Ciompi, L. 2004. 'Ein blinder Fleck bei Niklas Luhmann? Soziale Wirkungen von Emotionen aus Sicht der fraktalen Affektlogik', *Soziale Systeme. Zeitschrift für soziologische Theorie* 10 (1): 21–49.

Clark, C.W. 2005. *Mathematical Bioeconomics: The Optimal Management of Renewable Resources*, 2nd edn. New York: John Wiley & Sons.

Cramb, A. 1994. 'The Whale Debate is Back on the Agenda', *The Scotsman*, 10 June. Retrieved on 11 March 2010 from LexisNexis database.

Crossley, N. 2003. 'From Reproduction to Transformation: Social Movement Fields and the Radical Habitus', *Theory, Culture and Society* 20(6): 43–68.

Czarniawska, B. 2005. 'On Gorgon Sisters: Organizational Action in the Face of Paradox', in D. Seidl and K.H. Becker (eds), *Niklas Luhmann und Organization Studies*. Malmö: Liber & Copenhagen Business School Press, pp. 127–142.

Dale, S. 1996. *McLuhan's Children: The Greenpeace Message and the Media*. Toronto: Between the Lines.

Dalton, A. 1996. 'The Stock Questions', *The Scotsman*, 24 July, 12.

Darnton, J. 1993. 'Norwegians Claim Their Whaling Rights', *The New York Times*, 7 August, 1. Retrieved on 11 March 2010 from LexisNexis database.

Diamond, M.A. 1986. 'Resistance to Change: A Psychoanalytic Critique of Argyris and Schon's Contributions to Organization Theory and Intervention', *Journal of Management Studies* 23(5): 543–562.

———. 1993. *The Unconscious Life of Organizations: Interpreting Organizational Identity*. Westport and London: Quorum Books.

Directory of Fisheries. 2014. *Forskrift om regulering av fangst av vågehval i 2014*. Retrieved on 24 August 2014 from http://www.fiskeridir.no/fiske-og-fangst/j-meldinger/gjeldende-j-meldinger/j-90-2014.

Dopson, S., and J.E. Neumann. 1998. 'Uncertainty, Contrariness and the Double-bind: Middle Managers' Reactions to Changing Contracts', *British Journal of Management* 9(s1): S53–S70.
Dorsey, K. 2013. *Whales and Nations: Environmental Diplomacy on the High Seas*. Seattle and London: University of Washington Press.
Doyle, A. 2009. 'Norway Whale Catches Fall To Lowest In A Decade', *Thomson Reuters*, 28 August. Retrieved on 31 August 2009 from http://planetark.org/ark/54439.
Dutton, J.E., and J.M. Dukerich. 2004. 'Keeping an Eye on the Mirror: Image and Identity in Organizational Adaptation', in M.J. Hatch and M. Schultz (eds), *Organizational Identity. A Reader*. Oxford: Oxford University Press, pp. 183–222.
Eilam, G., and B. Shamir. 2005. 'Organizational Change and Self-Concept Threats. A Theoretical Perspective and a Case Study', *Journal of Applied Behavioral Science* 41(4): 399–421.
Ellemers, N. 2012. 'The Group Self', *Science* 336: 848–852.
———, and D. de Gilder. 2012. *Je werkt anders dan je denkt*. Amsterdam and Antwerpen: Uitgeverij Business Contact.
Emberland, T. 2000. 'Åpen båt hos Greenpeace', *Nordlys Morgen*, 18 May, 12.
Engesland, B.E., C. Guhnfeldt, E. Solvoll and H.O. Torgersen. 1992. 'Sterke reaksjoner på ny hvalfangst', *Aftenposten Morgen*, 30 June, 6.
Erwood, S. 2011. *The Greenpeace Chronicles*. Amsterdam: Greenpeace International.
Ford, J.D., and L.W. Ford. 1994. 'Logics of Identity, Contradiction, and Attraction in Change', *Academy of Management Review* 19(4): 756–785.
Fouché, G. 2008. 'Whaling Under Fire as Norway Catches Only 50% of Its Quota', *guardian.co.uk*, 22 August. Retrieved on 31 August 2009 from http://www.guardian.co.uk/environment/2008/aug/22/whaling.endangeredspecies.
Freud, A. 1967. *Jeg'et og forsvarsmekanismerne* [The Ego and the Mechanisms of Defence]. Copenhagen: Hans Reitzels Forlag A/S.
Freud, S. 1978. 'Remembering, Repeating and Working-Through', in J. Strachey (ed.), *The Standard Edition of the Complete Psychological Works of Sigmund Freud, Volume XII*. London: The Hogarth Press, pp. 145–156.
Furuly, J.G. 1993a. 'Hvalfangst-general sabotasjemål', *Aftenposten*, 4 March.
———. 1993b. '"Hvalkrigen" starter i april', *Aftenposten*, 4 March.
———. 1994. 'Skuta ble berget i siste liten', *Aftenposten Morgen*, 25 January, 2.
Gabriel, Y., and A. Carr. 2002. 'Organizations, Management and Psychoanalysis: An Overview', *Journal of Managerial Psychology* 17(5): 348–365.
Gamson, W.A. 1975. *The Strategy of Social Protest*. Homewood, IL: The Dorsey Press.
———. 1992. 'The Social Psychology of Collective Action', in A.D. Morris and C. McClurg Mueller (eds), *Frontiers in Social Movement Theory*. New Haven, CT: Yale University Press, pp. 53–76.

———. 2004. 'Bystanders, Public Opinion, and the Media', in D.A. Snow, S.A. Soule and H. Kriesi (eds), *The Blackwell Companion to Social Movements*. Malden, MA: Blackwell Publishing, pp. 242–261.

———, D. Croteau, W. Hoynes and T. Sasson. 1992. 'Media Images and the Social Construction of Reality', *Annual Review of Sociology* 18: 373–393.

Gans, H.J. 1979. *Deciding What's News. A Study of* CBS Evening News, NBC Nightly News, Newsweek, *and* Time. New York: Pantheon Books.

Geertz, C. 1973. 'Thick Description: Toward an Interpretive Theory of Culture', in C. Geertz (ed.), *The Interpretation of Cultures: Selected Essays*. New York: Basic Books, pp. 3–30.

Gitlin, T. 1980. *The Whole World Is Watching: Mass Media in the Making and Unmaking of the New Left*. Berkeley, Los Angeles and London: University of California Press.

Giugni, M.G. 1998. 'Was it Worth the Effort? The Outcomes and Consequences of Social Movements', *Annual Review of Sociology* 24: 371–393.

———. 1999. 'Introduction: How Social Movements Matter: Past Research, Present Problems, Future Developments', in M. Giugni, D. McAdam and C. Tilly (eds), *How Social Movements Matter*. Minneapolis, MN and London: University of Minnesota Press, pp. xiii–xxxiii.

Gjerde, A. 1986. 'Tidenes største folkefest', *NTBtekst,* 2 May.

Gjørvad, P. 1999. 'Ny, het hvalsommer?', *Dagens Næringsliv Morgen*, 8 July, 16.

Glosemeyer, L. 2000. 'greenpeace', *Nordlys Morgen*, 23 May, 16.

Goodman, A. with D. Goodman. 2004. *The Exception to the Rulers: Exposing Oily Politicians, War Profiteers, and the Media That Love Them*. New York: Hyperion.

Goodwin, J., J.M. Jasper and F. Polletta. 2000. 'The Return of the Repressed: The Fall and Rise of Emotions in Social Movement Theory', *Mobilization* 5(1): 65–83.

Gordon, H.S. 1954. 'The Economic Theory of a Common-Property Resource: The Fishery', *The Journal of Political Economy* LXII: 124–142.

Granaasen, I. 1985. 'Nye boikottaksjoner mot norsk hvalfangst', *Aftenposten Morgen*, 11 July, 8.

Greenpeace Nordic. 2014. *2013 Nordic* [Annual Report]. Stockholm: Greenpeace Nordic.

Grendstad, G., P. Selle, K. Strømsnes and Ø. Bortne. 2006. *Unique Environmentalism: A Comparative Perspective*. New York: Springer.

Grytås, G. 2014. 'Slutt for harpun og hakapik?', in N. Kolle (ed.), *Havet, fisken og oljen 1970–2014*. Bergen: Fagbokforlaget, pp. 317–341.

Günther, J. 2004. 'Organisation und Identität', in D. Baecker, F.E. Dievernich and T. Schmidt (eds), *Strategien der Organisation: Ressourcen - Strukturen - Kompetenzen*. Wiesbaden: Deutscher Universitäts-Verlag, pp. 1–38.

Haluza-DeLay, R. 2008. 'A Theory of Practice for Social Movements: Environmentalism and Ecological Habitus', *Mobilization* 13(2): 205–218.

Hansen, J.C., and A. Holt-Jensen. 1982. *Det moderne Norge: Ressursene våre.* Oslo: Gyldendal Norsk Forlag.

Hansen, B.E. 1994. '"Greenpeace-lov" vedtatt i går', *VG*, 16 July, 19.

Harbo, H. 1985. 'Hvalfangstkommisjonen har økonomiproblemer', *Aftenposten Morgen*, 19 July, 4.

Hare-Mustin, R.T. 1976. 'Paradoxical Tasks in Family Therapy: Who Can Resist?', *Psychotherapy: Theory, Research and Practice* 13(2): 128–130.

Hegna, L. 1986. 'Hvalfangere mot hangarskip', *Aftenposten Morgen*, 30 August, 8.

Helle, K.-E., and D. Stenerud. 2003. 'Spekk løgn & hvalfangst', *Folkevett* 6: 8–16.

Hellmann, K. 1998. 'Systemtheorie und Bewegungsforschung. Rezeptionsdefizite aufgrund von Stildifferenzen oder das Außerachtlassen von Naheliegendem', *Rechtshistorisches Journal* 17: 493–510.

———. 2000. '"… und ein größeres Stück Landschaft mit den erloschenen Vulkanen des Marxismus." Oder: Warum rezipiert die Bewegungsforschung Luhmann nicht?', in H. De Berg and J.F.K. Schmidt (eds), *Rezeption und Reflexion. Zur Resonanz der Systemtheorie Niklas Luhmanns außerhalb der Soziologie*. Frankfurt/Main: Suhrkamp Verlag, pp. 411–439.

Hennestad, B.W. 1990. 'The Symbolic Impact of Double Bind Leadership: Double Bind and the Dynamics of Organizational Culture', *Journal of Management Studies* 27(3): 265–280.

Hernes, T., and T. Bakken. 2003. 'Introduction: Niklas Luhmann's Autopoietic Theory and Organization Studies – a Space of Connection', in T. Bakken and T. Hernes (eds), *Autopoietic Organization Theory. Drawing on Luhmann's Social Systems Perspective*. Copenhagen: Abstrakt Liber Copenhagen Business School Press, pp. 9–22.

Hernes, H.-K., and K.H. Mikalsen. 2002. 'From Protest to Participation? Environmental Groups and the Management of Marine Fisheries', *Mobilization* 7(1): 15–28.

Hesstvedt, O. 1984. 'Felles kritikk av nedbyggingsstøtte: Allianse hvalfangere Greenpeace', *Aftenposten Morgen*, 24 May, 4.

Hesstvedt, K. 1992. '"Her" og "neri der"', *Dagens Næringsliv Morgen*, 26 August, 27.

———. 1993. 'Hvalfangst: Hva med IWCs troverdighet?', *Aftenposten Morgen*, 30 June, 15.

High North Alliance. 1997. 'Shift in Greenpeace Whaling Policy: May Accept Commercial Whaling', 27 May. Retrieved on 13 July 2011 from http://www.highnorth.no/news/pfriend.asp?which=121.

———. 2002. 'Whale Export Resumed', 15 July. Retrieved on 23 February 2012 from http://www.highnorth.no/news/nedit.asp?which=294.

———. (n.d.). David and Goliath (photo). Retrieved from http://www.highnorth.no/library/movements/Greenpeace/da-an-go.htm.

Hirschhorn, L., and T. Gilmore. 1980. 'The Application of Family Therapy Concepts to Influencing Organizational Behavior', *Administrative Science Quarterly* 25(1): 18–37.

Holm, P.A. 1983. 'Sjøslag om selfangst', *VG,* 18 March, 13.

Hunter, R. 1971. *The Storming of the Mind. Inside the Consciousness Revolution.* New York: Doubleday & Co.

———. 2011. *Warriors of the Rainbow. A Chronicle of the Greenpeace Movement from 1971 to 1979.* Amsterdam: Greenpeace/Fremantle Press.

Hveem, H., S. Lodgaard, K. Skjelsbaek and N.J. Ringdal. 1984. *Det moderne Norge: Vår plass i verden.* Oslo: Gyldendal Norsk Forlag.

Hvoslef, S. 1993. 'Miljøvernorganisasjonene og hvalfangsten – WWF', in N.C. Stenseth, A.H. Hoel and I.B. Lid (eds), *Vågehvalen – valgets kval.* Oslo: Ad Notam Gyldendal, pp. 173–182.

Icelandic Ministry of Fisheries and Agriculture. 2006. *Iceland and Whaling for Scientific Purposes.* Retrieved on 21 April 2011 from http://eng.sjavarutvegsraduneyti.is/news-and-articles/nr/1229.

———. 2009. *Continuation of Sustainable Whaling.* Retrieved on 21 April 2011 from http://eng.sjavarutvegsraduneyti.is/news-and-articles/nr/9554.

Jamison, A. 2004. 'Learning from Lomborg: Or Where Do Anti-Environmentalists Come From?', *Science as Culture* 13(2): 173–195.

———, R. Eyerman, J. Cramer, with J. Læssøe. 1990. *The Making of the New Environmental Consciousness. A Comparative Study of the Environmental Movements in Sweden, Denmark and the Netherlands.* Edinburgh: Edinburgh University Press.

Janis, I.L., and L. Mann. 1979. *Decision Making: A Psychological Analysis of Conflict, Choice, and Commitment.* New York: The Free Press.

Jentoft, W. 1989. 'Greenpeace vil stanse islandsk sel- og hvalfilm i NRK', *NTBtekst,* 2 April.

Johannessen, B.A., and R.S. Bertinussen. 1995. 'Det store Baste-taket', *VG,* 4 June, 30.

Johansen, H.P. 1999. 'Lovlig norsk hvalfangst', *Aftenposten Morgen,* 9 May, 10.

Jonassen, A.M. 1993. 'Den store HVALKRIGEN', *Aftenposten Morgen,* 7 April, 12.

Kahn, W.A. 2001. 'Holding Environments at Work', *Journal of Applied Behavioral Science* 37(3): 260–279.

———. 2003. 'The Revelation of Organizational Trauma', *Journal of Applied Behavioral Science* 39(4): 364–380.

———. 2004. 'Facilitating and Undermining Organizational Change', *Journal of Applied Behavioral Science* 40(1): 7–30.

Kalland, A. 1993. 'Whale Politics and Green Legitimacy: A Critique of the Anti-Whaling Campaign', *Anthropology Today* 9(6): 3–7.

———. 2012. *Unveiling the Whale: Discourses on Whales and Whaling.* New York and Oxford: Berghahn Books.

Kets de Vries, M.F.R., and D. Miller. 1984. *The Neurotic Organization*. San Francisco, Washington and London: Jossey-Bass.

Kihlberg, D. 2011. *Motstånd i skogen: En motiv- og maktanalys av civila olydnadsaktioner i svensk skogspolitik*. Kandidatuppsats i Statsvetenskap/ Miljöpolitik, Göteborgs Universitet. Retrieved on 23 November 2015 from https://gupea.ub.gu.se/bitstream/2077/27951/1/gupea_2077_27951_1.pdf.

Kimura, T. 2014. 'Newspaper Reporting of Whaling in Australia and Japan: A Comparative Content Analysis', *New Voices* 6: 173–198.

King, M. 1986. *Death of the Rainbow Warrior*. Auckland: Penguin Books.

Klokk, A.M. 1978. 'Aksjon: "Redd hvalen"', *VG*, 27 May, 13.

Koch, S. 1999. *Höger om! En svensk historia 1968–98*. Stockholm: Ordfront.

Kotter, J.P. 1995. 'Leading Change: Why Transformation Efforts Fail', *Harvard Business Review* March-April: 59–67.

Kramer, J. 1984. 'Norsk identitet - et produkt av underutvikling og stammetilhørighet', in A. M. Klausen (ed.), *Den norske væremåten: Antropologisk søkelys på norsk kultur*. Oslo: J. W. Cappelens Forlag AS, pp. 88–97.

Krantz, J. 1999. 'Comment on "Challenging 'Resistance to Change'"', *Journal of Applied Behavioral Science* 35(1): 42–44.

Kristiansen, S.K. 1984. 'Greenpeace nådeløse i USA: - Boikott alt norsk', *VG*, 13 January, 15.

Kristjánsson, M. 2015. 'Hvalfangst', *Klassekampen*, 2 July, 2.

Krog, K., and S. Talsnes. 1992. 'Norge i mediestorm', *VG*, 30 June, 6.

Kvistad, Y., S. Frafjord and T. Bøe. 1996. 'Raser mot Norge', *VG*, 15 May, 1.

Laing, R.D. 1976. 'Mystification, Confusion, and Conflict (1965)', in C.E. Sluzki and D.C. Ransom (eds), *Double Bind: The Foundation of the Communicational Approach to the Family*. New York, London and San Francisco: Grune & Stratton, pp. 199–218.

Lakoff, G. 2004. *Don't Think of an Elephant! Know Your Values and Frame the Debate*. White River Junction, VT: Chelsea Green Publishing.

Leathers, D.G. 1979. 'The Impact of Multichannel Message Inconsistency on Verbal and Nonverbal Decoding Behaviors', *Communication Monographs* 46(2): 88–100.

Livesey, S.M. 2001. 'Eco-Identity as Discursive Struggle: Royal Dutch/Shell, Brent Spar, and Nigeria', *Journal of Business Communication* 38(1): 58–91.

Lloyd-Roberts, S. 1991. 'Endangered Species Nobody Wants to Save; Whalers are Demanding the Right to Work', *The Independent*, 27 May, 15. Retrieved on 11 March 2010 from LexisNexis database.

Louis, W.R. 2009. 'Collective Action- and Then What?', *Journal of Social Issues* 65(4): 727–748.

Luhmann, N. 1984. *Soziale Systeme: Grundriß einer allgemeinen Theorie*. Frankfurt/Main: Suhrkamp.

———. 1988. 'Njet-Set und Terror-Desperados', *taz*, 4 August, pp. 11–12. Retrieved on 2 October 2015 from http://organisationen.freepage.de/cgi-bin/feets/freepage_ext/41030x030A/rewrite/blk/BLK2/blk2-16.htm.

———. 1990. *Ökologische Kommunikation: Kann die moderne Gesellschaft sich auf ökologische Gefährdungen einstellen?*, 3. Aufl. Opladen: Westdeutscher Verlag.

———. 1995a. 'Konzeptkunst. Brent Spar oder Können Unternehmen von der Öffentlichkeit lernen?', *Frankfurter Allgemeine Zeitung*, 19 July.

———. 1995b. *Social Systems*, trans. J. Bednarz, Jr., with D. Baecker. Stanford, CA: Stanford University Press.

———. 1997a. 'Frauen, Männer und George Spencer Brown', in N. Luhmann, *Protest. Systemtheorie und soziale Bewegungen.* Hrsg. und eingeleitet von K.-U. Hellmann, 2. Aufl. Frankfurt/Main: Suhrkamp Verlag, pp. 107–155.

———. 1997b. *Die Gesellschaft der Gesellschaft.* Frankfurt/Main: Suhrkamp.

———. 1997c. *Protest. Systemtheorie und soziale Bewegungen.* Hrsg. und eingeleitet von K.-U. Hellmann, 2. Aufl. Frankfurt/Main: Suhrkamp.

———. 2006. *Organisation und Entscheidung*, 2. Aufl. Wiesbaden: VS Verlag für Sozialwissenschaften.

Lysvold, S. 2014. 'Greenpeace: -Hvalnæringa vil dø ut av seg selv', *NRK*, 18 May. Retrieved on 24 August 2014 from http://www.nrk.no/nordland/_-hvalnaeringa-vil-forsvinne-1.11714668.

Maddox, B. 1992. 'Return of the Whale Killers: Norway Defends the Hunt as "Cultural Preservation"', *The Financial Post*, 3 August, 46. Retrieved on 11 March 2010 from LexisNexis database.

Mansbridge, J.J. 1986. *Why We Lost the ERA.* Chicago and London: University of Chicago Press.

March, J.G., and J.P. Olsen. 1976. *Ambiguity and Choice in Organizations*, 2nd edn. Bergen, Oslo and Tromsø: Universitetsforlaget.

March, J.G., and H.A. Simon. 1958. *Organizations.* New York: John Wiley & Sons, Inc.

Mathisen, M.C. 1999. 'Greenpeace til Tromsø', *Nordlys Morgen*, 8 May, 5.

Mathismoen, O. 1992a. 'Krigen Norge vil tape', *Aftenposten Morgen*, 8 August, 22.

———. 1992b. 'Mange bomskudd på hval', *Aftenposten Morgen*, 14 November, 4.

———, and A.M. Jonassen. 1993. 'Kamp om inntekt og hvalen som symbol', *Aftenposten Morgen*, 7 April, 13.

Maturana, H.R. 1999. 'The Organization of the Living: A Theory of the Living Organization', *International Journal of Human-Computer Studies* 51: 149–168.

——— and F.J. Varela. 1987. *Der Baum der Erkenntnis: Die biologischen Wurzeln menschlichen Erkennens*, 11. Aufl. Munich: Goldmann.

McAdam, D., J.D. McCarthy and M.N. Zald. 1988. 'Social Movements', in N.J. Smelser (ed.), *Handbook of Sociology*. Newbury Park, CA: Sage, pp. 695–737.

McLuhan, E. (n.d.). 'The Source of the Term, "Global Village"', *McLuhan Studies*. Retrieved on 7 November 2014 from http://projects.chass.utoronto.ca/mcluhan-studies/v1_iss2/1_2art2.htm.

McLuhan, M. and Q. Fiore. 2001. *The Medium is the Massage. An Inventory of Effects*. Corte Madera, CA: Gingko Press.

McTaggart, D.F. 1973. *Outrage! The Ordeal of Greenpeace III*. Vancouver: J. J. Douglas Ltd.

Medvetz, T. 2006. 'The Strength of Weekly Ties: Relations of Material and Symbolic Exchange in the Conservative Movement', *Politics & Society* 34(3): 343–368.

Meyer, J.W., and B. Rowan. 1977. 'Institutionalized Organizations: Formal Structure as Myth and Ceremony', *American Journal of Sociology* 83(2): 340–363.

Mikalsen, Ø. 1999a. 'Hval-kollisjon', *Dagbladet*, 13 June, 12.

———. 1999b. 'Skjøt ikke for å treffe: Bastesen forsvarer skudd mot hvalfangstdemonstranter', *Dagbladet*, 13 July, 14.

———, and M. Schjølberg Ulshagen. 1999. 'Greenpeace fikk servert hval-pizza', *Dagbladet*, 16 June, 18.

Minkoff, D.C., and J.D. McCarthy. 2005. 'Reinvigorating the Study of Organizational Processes in Social Movements', *Mobilization* 10(2): 289–308.

Motavalli, J. 1995. 'Going For the Jugular – Greenpeace Critics', *E: The Environmental Magazine*, November-December. Retrieved on 2 August 2007 from www.findarticles.com/p/articles/mi_m1594/is_n6_v6/ai_17847913.

Murata, K. 2007. 'Pro- and Anti-Whaling Discourses in British and Japanese Newspaper Reports in Comparison: A Cross-Cultural Perspective', *Discourse & Society* 18(6): 741–764.

Murphy, P., and J. Dee. 1992. 'Du Pont and Greenpeace: The Dynamics of Conflict Between Corporations and Activist Groups', *Journal of Public Relations Research* 4(1): 3–20.

Myklebust, M. 1993. 'Portrett', *Nordlys Morgen*, 24 April, 32.

Nag, R., K.G. Corley and D.A. Gioia. 2007. 'The Intersection of Organizational Identity, Knowledge, and Practice: Attempting Strategic Change via Knowledge Grafting', *Academy of Management Journal* 50(4): 821–847.

Nassehi, A., and G. Nollmann. 2004. 'Einleitung: Wozu ein Theorienvergleich?', in A. Nassehi and G. Nollmann (eds), *Bourdieu und Luhmann: Ein Theorienvergleich*. Frankfurt/Main: Suhrkamp, pp. 7–22.

Nielsen, B. 1992. '– Miljø-fascisme', *VG*, 26 July, 9.

———, and R. Stoltz Bertinussen. 1993. 'Hjemmekoselig aksjon', *VG*, 27 May, 23.

N. N. 1984. 'Ingen tør hente dem ned!', *Aftenposten Aften*, 30 October, 1.
N. N. 1986a. 'Carrier Nimitz Met By Protesting Norwegian Whalers', *The Associated Press*, 29 August. Retrieved on 11 March 2010 from LexisNexis database.
N. N. 1986b. 'Greenpeace-fartøy oppbrakt og ført til Vardø', *NTBtekst*, 12 June.
N. N. 1986c. 'Greenpeace til sjøs i utkikkstønne', *NTBtekst*, 2 July.
N. N. 1986d. '"Moby Dick" til Oslo, håper å møte regjeringsmedlemmer', *NTBtekst*, 7 July.
N. N. 1986e. 'Greenpeaceskipet: "Moby Dick" forlater Oslo', *Aftenposten Aften*, 14 July, 7.
N. N. 1986f. 'Greenpeace-protest mot Norge', *NTBtekst*, 9 June.
N. N. 1990. 'Greenpeace invaderer kyst-Norge', *NTBtekst*, 4 July.
N. N. 1991. 'Greenpeace: - Prøve før kommersiell fangst', *NTBtekst*, 2 December.
N. N. 1992. 'Hvalfangst – en gave til miljøbevegelsene', *Aftenposten Morgen*, 31 December, 12.
N. N. 1994a. '24 000 "blodige" postkort', *Aftenposten Morgen*, 24 May, 6.
N. N. 1994b. 'Ny aksjon fra Greenpeace', *Aftenposten Morgen*, 7 July, 4.
N. N. 1994c. 'Landet rundt Greenpeace-terror mot hvalfangere', *Aftenposten Morgen*, 14 July, 2.
N. N. 1994d. 'Nytt våpen', *Aftenposten Morgen*, 19 July, 6.
N. N. 1994e. '"Solo"-skipperen bøtelagt', *Aftenposten Morgen*, 30 July, 4.
N. N. 1995a. 'Whaler Scuffles with Protesters', *The Toronto Star*, 4 June, A16. Retrieved on 11 March 2010 from LexisNexis database.
N. N. 1995b. 'Greenpeace-erstatning til Senet-skipper', *NTBtekst*, 11 May.
N. N. 1995c. 'Jacht op Greenpeace', *De Groene Amsterdammer*, 28 June. Retrieved on 30 March 2016 from https://www.groene.nl/artikel/jacht-op-greenpeace.
N. N. 1996a. 'Greenpeace Ship Sails into Firth to Confront Danes', *The Herald*, 7 May, 4.
N. N. 1996b. 'Watson tar ansvaret for angrep på norsk hvalskute', *NTBtekst*, 15 May.
N. N. 1996c. 'Greenpeace avbrøt aksjon mot hvalfangere', *NTBtekst*, 14 June.
N. N. 1996d. 'Fisheries: EC Slashes Herring Quotas in Half', *Greenwire*, 3 July.
N. N. 1996e. 'Untitled', *Associated Press Worldstream*, 4 July.
N. N. 1997. 'Etterlyser mann etter aksjonen mot Senet', *NTBtekst*, 1 May.
N. N. 1999a. 'Greenpeace-aktivister løslatt', *NTBtekst*, 14 June.
N. N. 1999b. 'Slakter Norge', *Dagbladet*, 08 July, 19.
N. N. 1999c. 'Hvalskuddene', *Dagens Næringsliv Morgen*, 15 July, 2.
N. N. 1999d. 'Hvalskipper inviterer Greenpeace om bord', *Aftenposten Morgen*, 17 July, 5.
N. N. 2000. 'Greenpeace endrer strategi i kampen mot hvalfangst', *NTBtekst*, 10 May.
N. N. 2001. 'Hvalfanger anker ikke dom', *NTBtekst*, 6 January.

Nordeng, T. 1984. 'Norge har ingen hval å eksportere', *Aftenposten Morgen*, 29 February, 11.

Norsk Gallup Institutt AS. 2002. *Norsk natur- og miljøbarometer*. Copy obtained from Greenpeace Norway.

Norsk Kulturråd. 2011. *Hval, veid, fangst og norske kyster: Linjer i norsk hvalfangsthistorie*. Retrieved on 24 September 2015 from http://hvalfangstarkiver.no/wp-content/uploads/2014/01/linjer-i-norsk-hvalfangsthistorie.pdf.

Norwegian Ministry of Fisheries and Coastal Affairs. 2004. *Kvote for fangst av vågekval i 2005*. Retrieved on 14 March 2007 from http://www.regjeringen.no/nb/dokumentarkiv/Regjeringen-Bondevik-II/fkd/Nyheter-og-pressemeldinger/2004/kvote_for_fangst_av_vagekval_i.html?id=253050.

———. 2005. *Kvote for fangst av vågekval i 2006*. Retrieved on 14 March 2007 from http://www.regjeringen.no/nn/dep/fkd/pressesenter/pressemeldinger/2005/kvote-for-fangst-av-vagekval-i-2006.html?id=103519.

———. 2006. *Kvote for fangst av vågekval i 2007*. Retrieved on 5 March 2012 from http://www.regjeringen.no/nn/dep/fkd/pressesenter/Pressemeldingar/2006/kvote-for-fangst-av-vagekval-i-2007.html?id=436844.

———. 2008. *Kvote for fangst av vågekval i 2008*. Retrieved on 5 March 2012 from http://www.regjeringen.no/nn/dep/fkd/pressesenter/Pressemeldingar/2008/kvote-for-fangst-av-vagekval-i-2008.html?id=499375.

———. 2010. *Kvote for fangst av vågekval i 2011*. Retrieved on 5 March 2012 from http://www.regjeringen.no/nn/dep/fkd/pressesenter/Pressemeldingar/2010/kvote-for-fangst-av-vagekval-i-2011.html?id=629357.

———. 2013. *Kvote for fangst av vågehval i 2013*. Retrieved on 24 August 2014 from http://www.regjeringen.no/nb/dokumentarkiv/stoltenberg-ii/fkd/Nyheter_og_pressemeldinger/Pressemeldinger/2013/kvote-for-fangst-av-vagehval-i-2013.html?regj_oss=1&id=713911.

Norwegian Ministry of Foreign Affairs. 2003. *Der Fang von Zwergwalen in Norwegen*. Retrieved on 12 July 2006 from http://www.norwegen.no/misc/print.aspx?article={4ac13188-5cac-44c9-99bd-0a1576ee0f48}.

Oakley, J.G. 2000. 'Gender-Based Barriers to Senior Management Positions: Understanding the Scarcity of Female CEOs', *Journal of Business Ethics* 27(4): 321–334.

Olaussen, L.M. 1999. 'Bastesen angrer ikke skytingen', *Aftenposten Morgen*, 14 July, 4.

Olsen, E. 1991. 'Sanksjoner og boikott truer', *VG*, 2 November, 3.

———. 1994. 'Kjemper sammen', *VG*, 11 May, 14–16.

Olson, D.H. 1972. 'Empirically Unbinding the Double Bind: Review of Research and Conceptual Reformulations', *Family Process* 11(1): 69–94.

Omdahl, J. 1993a. 'Hval på hjernen', *Dagens Næringsliv Morgen*, 22 May, 28.

———. 1993b. 'Til felts mot Greenpeace', *Dagens Næringsliv Morgen*, 5 June, 8.

Papp, P. 1980. 'The Greek Chorus and Other Techniques of Paradoxical Therapy', *Family Process* 19: 45–57.
Pleym, F. 1999a. 'Hvalfangst og Greenpeace', *Dagens Næringsliv Morgen*, 27 July, 3.
———. 1999b. 'Blåhval-biff på det japanske markedet', *Dagens Næringsliv Morgen*, 23 August, 4.
———. 1999c. 'Norge ignorerer IWC', *Nordlys Morgen*, 25 August, 14.
———. 2000. 'Mer enn vågehval', *Dagens Næringsliv Morgen*, 31 July, 3.
Popovic, S., and M. Miller. 2015. *Protest! Wie man die Mächtigen das Fürchten lehrt*. Frankfurt/Main: S. Fischer Verlag.
Putnam, L.L. 1986. 'Contradictions and Paradoxes in Organizations', in L. Thayer (ed.), *Organization – Communication. Emerging Perspectives I*. Norwood, NJ: Ablex Publishing Corporation, pp. 151–167.
Ramberg, E., and A. Henriksen. 1994. 'Slo hardt ned på protestmur', *Aftenposten Morgen*, 23 July, 15.
Randsborg, E. 1996. 'Bølger i Greenpeace, dansk leder må gå', *Aftenposten Morgen*, 10 July, 6.
Rapp, O.M. 1996. 'Hvalskute ble forsøkt senket', *Aftenposten Morgen*, 4 May, 2.
Rasmussen, A.R. 1986. 'Hvorfor fanger Norge hval? Hvalen og moralen', *Aftenposten Morgen*, 30 December, 9.
Rasmussen, L.T. 1989. 'Greenpeace til sak mot filmprodusent', *NTBtekst*, 24 May.
Reger, R.K., L.T. Gustafson, S.M. Demarie and J.V. Mullane. 1994. 'Reframing the Organization: Why Implementing Total Quality is Easier Said than Done', *Academy of Management Review* 19(3): 565–584.
Riese, J. 2011. 'Functions, Communication, and Perception of Emotions in Luhmannian Theory: Emotions as Reflection Resources of Social Systems', *Soziale Systeme. Zeitschrift für soziologische Theorie* 17(1): 53–72.
———. 2014. 'On Dynamic Processes of Framing, Counterframing, and Reframing: The Case of the Greenpeace Whale Campaign in Norway', in K. Fahlenbrach, E. Sivertsen and R. Werenskjold (eds), *Media and Revolt. Strategies and Performances from the 1960s to the Present*. New York and Oxford: Berghahn Books, pp. 283–299.
———. 2015. 'Taking Arms Against a Sea of Troubles: The Greenpeace Whale Campaign in Norway', *Journal of Applied Behavioral Science* 51(1): 94–128.
Rose, C. 1998. *The Turning of the 'Spar*. London: Greenpeace.
Rucht, D., and R. Roth. 1992. '"Über den Wolken…" Niklas Luhmanns Sicht auf soziale Bewegungen', *Forschungsjournal Neue Soziale Bewegungen* 5(2): 22–33.
Ruud, K. 1983. 'Kampen for vern av hval', *Aftenposten Morgen*, 2 November, 17.
Schein, E.H. 2006. 'From Brainwashing to Organizational Therapy: A Conceptual and Empirical Journey in Search of "Systemic" Health and a

General Model of Change Dynamics. A Drama in Five Acts', *Organization Studies* 27(2): 287–301.

Schiefloe, I. 1993. 'Greenpeace har satt i gang omfattende boikott', *Dagens Næringsliv Morgen*, 28 May, 15.

Schmidt, N. 1992. 'Nordmenn er spyttslikkere', *VG*, 24 July, 5.

Schwartz, H.S. 2004. 'Anti-Social Actions of Committed Organizational Participants: An Existential Psychoanalytic Perspective', in M.J. Hatch and M. Schultz (eds), *Organizational Identity: A Reader*. Oxford and New York: Oxford University Press, pp. 119–133. Originally published as Schwartz, H.S. 1987. 'Anti-Social Actions of Committed Organizational Participants: An Existential Psychoanalytic Perspective', *Organization Studies* 8: 327–340.

Seidl, D. 2003. 'Organisational Identity in Luhmann's Theory of Social Systems', in T. Bakken and T. Hernes (eds), *Autopoietic Organization Theory. Drawing on Luhmann's Social Systems Perspective*. Copenhagen: Abstrakt Liber Copenhagen Business School Press, pp. 123–150.

———, and K.H. Becker. 2005a. 'Introduction: Luhmann's Organization Theory', in D. Seidl and K.H. Becker (eds), *Niklas Luhmann und Organization Studies*. Malmö: Liber & Copenhagen Business School Press, pp. 8–18.

———, and K.H. Becker (eds). 2005b. *Niklas Luhmann und Organization Studies*. Malmö: Liber & Copenhagen Business School Press.

Shoham-Salomon, V., R. Avner and R. Neeman. 1989. 'You're Changed If You Do and Changed If You Don't: Mechanisms Underlying Paradoxical Interventions', *Journal of Consulting and Clinical Psychology* 57(5): 590–598.

Simon, F.B. 2000a. *Meine Psychose, mein Fahrrad und ich. Zur Selbstorganisation der Verrücktheit*, 8. Aufl. Heidelberg: Carl-Auer-Systeme Verlag.

———. 2000b. 'Name Dropping. Zur erstaunlich großen, bemerkenswert geringen Rezeption Luhmanns in der Familienforschung', in H. De Berg and J.F.K. Schmidt (eds), *Rezeption und Reflexion. Zur Resonanz der Systemtheorie Niklas Luhmanns außerhalb der Soziologie*. Frankfurt/Main: Suhrkamp, pp. 361–386.

———. 2004. 'Zur Systemtheorie der Emotionen', *Soziale Systeme. Zeitschrift für soziologische Theorie* 10(1): 111–139.

Skjalg, F. 1999. 'Motstridende påstander om hval-aksjon i Nordsjøen', *Aftenposten Morgen*, 13 July, 4.

———. 2000. 'Betinget fengsel for skudd mot Greenpeace', *NTBtekst*, 23 November.

Sluzki, C.E., J. Beavin, A. Tarnopolsky and E. Veron. 1967. 'Transactional Disqualification: Research on the Double Bind', *Archives of General Psychiatry* 16: 494–504.

Smith, K.K., and D.N. Berg. 1987. *Paradoxes of Group Life. Understanding Conflict, Paralysis, and Movement in Group Dynamics*. San Francisco and London: Jossey-Bass.

Soldow, G.F. 1981. 'Change in the Organization: The Detriment and Benefit of the Double Bind', *Group & Organization Studies* 6(4): 500–513.

Sollied, K.-A. 1994. 'Hval: Grave en grav', *Nordlys Morgen*, 15 December, 21.

Spence, T. 1996. '–IWC-møtet suksess for Norge', *Aftenposten Morgen*, 27 June, 4.

Stake, R.E. 1995. *The Art Of Case Study Research*. Thousand Oaks, London and New Delhi: Sage.

Statistics Norway. 2006. *Kvalfangst*. Retrieved on 21 April 2011 from http://www.ssb.no/aarbok/2006/fig/fig-375.html.

———. 2010. *Småhvalfangsten*. Retrieved on 21 April 2011 from http://www.ssb.no/histstat/tabeller/15-7.html.

Stein, H.F. 2001. *Nothing Personal, Just Business: A Guided Journey Into Organizational Darkness*. Westport: Quorum Books.

Stenerud, D. 2006. 'Skulle bade OL i "hvalblod"', *Nettavisen*. Retrieved on 20 June 2013 from www.na24.no/article737842.ece.

Stenseth, N.C., I.B. Lid and A.H. Hoel. 1993. 'Vågehvalfangsten: Den historiske utviklingen', in N.C. Stenseth, A.H. Hoel and I.B. Lid (eds), *Vågehvalen – valgets kval*. Oslo: Ad Notam Gyldendal, pp. 77–91.

Stewart, W.A. 1963. 'An Inquiry Into the Concept of Working Through', *Journal of the American Psychoanalytic Association* 11: 474–499.

Strand, T. 1985. 'Skal banne i kjerka', *VG*, 15 June, 27.

Straume, K.E. 1992. 'Hvalfangernes skrek', *Nordlys Morgen*, 21 July, 9.

Strømsnes, K., P. Selle and G. Grendstad. 2009. 'Environmentalism Between State and Local Community: Why Greenpeace Has Failed in Norway', *Environmental Politics* 18(3): 391–407.

Svarstad, A. 1982. 'VG drukner i postkort: Selsom aksjon', *VG*, 13 February.

Sætra, G. 1995. 'Dramatisk Greenpeace-aksjon mot Bastesen', *NTBtekst*, 3 June.

———. 1997. 'Nynyny!!!', *Nordlys Morgen*, 21 May, 25.

Sørensen, H. 1993. 'Miljøbevegelsen og vågehvalfangst – Natur og ungdom', in N.C. Stenseth, A.H. Hoel and I.B. Lid (eds), *Vågehvalen – valgets kval*. Oslo: Ad Notam Gyldendal, pp. 183–186.

Sørhaug, H.C. 1984. 'Totemisme på norsk: betraktninger om den norske sosialdemokratismes vesen', in A.M. Klausen (ed.), *Den norske væremåten: Antropologisk søkelys på norsk kultur*. Oslo: J. W. Cappelens Forlag, pp. 61–87.

Tamnes, R. 1997. *Oljealder 1965–1995*. Oslo: Universitetsforlaget.

Tilly, C. 1999. 'Conclusion: From Interactions to Outcomes in Social Movements', in M. Giugni, D. McAdam and C. Tilly (eds), *How Social Movements Matter*. Minneapolis, MN and London: University of Minnesota Press, pp. 253–270.

———. 2006. 'Afterword: Political Ethnography as Art and Science', *Qualitative Sociology* 29(3): 409–412.

Tonstad, P.L. 1999. 'Skyter hval ulovlig Norsk gavepakke til Greenpeace', *Dagbladet*, 11 July, 16.

Tracy, S.J. 2004. 'Dialectic, Contradiction, or Double Bind? Analyzing and Theorizing Employee Reactions to Organizational Tension', *Journal of Applied Communication Research* 32(2): 119–146.

Tømmerås, O. 1992. 'PR-krig på havet', *VG*, 28 July, 9.

Valenstein, A.F. 1983. 'Working Through and Resistance to Change: Insight and the Action System', *Journal of the American Psychoanalytic Association* 31S: 353–373.

Van Buskirk, W., and D. McGrath. 1999. 'Organizational Cultures as Holding Environments: A Psychodynamic Look at Organizational Symbolism', *Human Relations* 52(6): 805–832.

Van Zomeren, M., R. Spears and C.W. Leach. 2008. 'Exploring Psychological Mechanisms of Collective Action: Does Relevance of Group Identity Influence How People Cope With Collective Disadvantage?', *British Journal of Social Psychology* 47(2): 353–372.

Vaughan, A. 2010. 'Whale Cull Plan Sunk As National Delegates Fail to Agree', *guardian.co.uk*, 23 June. Retrieved on 15 July 2011 from http://www.guardian.co.uk/environment/2010/jun/23/whale-cull-plan-talks-fall-apart/print.

Veigård, E. 1999. 'Greenpeace skriver til tusener av nordmenn', *NTBtekst*, 31 May.

Vidal, J. 2011. 'Whaling Body Outlaws Malpractice With Anti-Corruption Reform', *guardian.co.uk*, 14 July. Retrieved on 15 July 2011 from http://www.guardian.co.uk/environment/2011/jul/14/whaling-body-anti-corruption/print.

———, and D. Adam. 2010. 'Return to Commercial Whaling Could Be Determined By Tiny Pacific Island', *guardian.co.uk*, 18 June. Retrieved on 15 July 2011 from http://www.guardian.co.uk/environment/2010/jun/18/return-to-commercial-whaling/print.

Vince, R., and M. Broussine. 1996. 'Paradox, Defense and Attachment: Accessing and Working With Emotions and Relations Underlying Organizational Change', *Organization Studies* 17(1): 1–21.

Vos, J.-P. 2005. 'Strategic Management from a Systems-Theoretical Perspective', in D. Seidl and K.H. Becker (eds), *Niklas Luhmann and Organization Studies*. Malmö. Liber & Copenhagen Business School Press, pp. 365–385.

Waagbø, A.J. 1992. 'Greenpeace-sjef går av i protest', *Dagbladet*, 2 August, 11.

Wagner, J.A. 1978. 'The Organizational Double Bind: Toward an Understanding of Rationality and Its Complement', *Academy of Management Review* 3(4): 786–795.

Wallace, A. 1992. 'Norway Plans to Resume Whale Hunting in 1993', *United Press International*, 29 June. Retrieved on 11 March 2010 from LexisNexis database.

Wasser, H. 2003. *Luhmanns Theorie psychischer Systeme und das Freudsche Unbewusste: Zur Beobachtung strukturfunktionaler Latenz.* Retrieved on 10 December 2014 from http://sammelpunkt.philo.at:8080/898/1/w1.pdf, V.1.

Watzlawick, P. 1978. *The Language of Change: Elements of Therapeutic Communication.* New York: Basic Books.

———, J. Helmick Beavin and D.D. Jackson. 1968. *Pragmatics of Human Communication: A Study of Interactional Patterns, Pathologies, and Paradoxes.* London: Faber and Faber.

———, J.H. Weakland and R. Fisch. 1974. *Change. Principles of Problem Formation and Problem Resolution.* New York and London: W.W. Norton & Co.

Weinbach, C. 2004a. 'Systemtheorie und Gender: Geschlechtliche Ungleichheit in der funktional differenzierten Gesellschaft', in S. Kampmann, A. Karentzos and T. Küpper (eds), *Gender Studies und Systemtheorie. Studien zu einem Theorietransfer.* Bielefeld: transcript Verlag, pp. 47–76.

———. 2004b. '… und gemeinsam zeugen sie geistige Kinder: Erotische Phantasien um Niklas Luhmann und Pierre Bourdieu', in A. Nassehi and G. Nollmann (eds), *Bourdieu und Luhmann. Ein Theorienvergleich.* Frankfurt am Main: Suhrkamp Verlag, pp. 57–84.

Westengen, K. 1994. 'Greenpeace forsøkte å stjele kanonen på Senet', *NTBtekst*, 2 July.

———. 1999. 'Greenpeace videofilmet mens "Villduen" tok hval', *NTBtekst*, 13 June.

Weyler, R. 2004. *Greenpeace. How a Group of Journalists, Ecologists and Visionaries Changed the World.* London: Rodale.

Whale and Dolphin Conservation Society. Not dated. *Subsidies in the Norwegian Whaling Industry.* Retrieved on 31 May 2006 from http://www.wdcs.org/dan/publishing.nsf/c525f7df6cbf01ef802569d600573108/4bfbf08e3461f1d2802568f10038ab85/$FILE/subs.pdf.

Wiedswang, K. 1993. 'Kåre får hvalebrokk', *Dagens Næringsliv Morgen*, 11 May, 36.

Wilden, A. 1972. *System and Structure: Essays in Communication and Exchange.* London: Tavistock.

Wilder, C. 1979. 'The Palo Alto Group: Difficulties and Directions of the Interactional View for Human Communication Research', *Human Communication Research* 5(2): 171–186.

Willke, H. 1999. *Systemtheorie II: Interventionstheorie. Grundzüge einer Theorie der Intervention in komplexe Systeme,* 3. Aufl. Stuttgart: Lucius & Lucius.

Winnicott, D.W. 1960. 'The Theory of the Parent-Infant Relationship', *The International Journal of Psychoanalysis* 41: 585–595.

———. 1990 [1963]. 'Psychiatric Disorder in Terms of Infantile Maturational Processes', in D.W. Winnicott (ed.), *The Maturational Processes and the*

Facilitating Environment. London: Karnac Books and the Institute of Psycho-Analysis, pp. 230–241.
Wynne, L.C. 1976. 'On the Anguish, and Creative Passions, of Not Escaping Double Binds: A Reformulation', in C.E. Sluzki and D.C. Ransom (eds), *Double Bind: The Foundation of the Communicational Approach to the Family*. New York, London and San Francisco: Grune & Stratton, pp. 243–250.
Yin, R.K. 1994. *Case Study Research: Design and Methods*, 2nd edn. Thousand Oaks, London and New Delhi: Sage.
Yngland, D. 1992. 'Gro og hvalen', *Dagens Næringsliv Morgen*, 6 August, 2.
Zangl, B. 1999. *Interessen auf zwei Ebenen: Internationale Regime in der Agrarhandels-, Währungs- und Walfangpolitik*. Baden-Baden: Nomos.
Zelko, F. 2007. 'Greenpeace and the Development of International Environmental Activism in the 1970s', in U. Lehmkuhl and H. Wellenreuther (eds), *Historians and Nature. Comparative Approaches to Environmental History*. Oxford and New York: Berg, pp. 296–318.
———. 2012. 'From Blubber and Baleen to Buddha of the Deep: The Rise of the Metaphysical Whale', *Society & Animals* 20: 91–108.
———. 2013. *Make It a Green Peace! The Rise of Countercultural Environmentalism*. Oxford: Oxford University Press.
Zinn, H. 2009 [2005]. 'Changing Minds, One at a Time', in H. Zinn, *The Zinn Reader: Writings on Disobedience and Democracy*. New York: Seven Stories Press, 2nd edn., pp. 693–697.
Øen, E.O. 1993. 'Fangst av vågehval i dag', in N.C. Stenseth, A.H. Hoel and I.B. Lid (eds), *Vågehvalen – valgets kval*. Oslo: Ad Notam Gyldendal, pp. 93–98.

Websites

High North Alliance: www.highnorth.no (expired)
International Union for Conservation of Nature: www.iucnredlist.org
Norwegian government: www.regjeringen.no
Statistics Norway: www.ssb.no

Index

Protest, Culture and Society

General editors:
Kathrin Fahlenbrach, Institute for Media and Communication, University of Hamburg
Martin Klimke, New York University Abu Dhabi
Joachim Scharloth, Technische Universität Dresden

Protest movements have been recognized as significant contributors to processes of political participation and transformations of culture and value systems, as well as to the development of both a national and transnational civil society.

This series brings together the various innovative approaches to phenomena of social change, protest and dissent which have emerged in recent years, from an interdisciplinary perspective. It contextualizes social protest and cultures of dissent in larger political processes and socio-cultural transformations by examining the influence of historical trajectories and the response of various segments of society, political and legal institutions on a national and international level. In doing so, the series offers a more comprehensive and multi-dimensional view of historical and cultural change in the twentieth and twenty-first centuries.

Volume 1
Voices of the Valley, Voices of the Straits: How Protest Creates Communities
Donatella della Porta and Gianni Piazza

Volume 2
Transformations and Crises: The Left and the Nation in Denmark and Sweden, 1956–1980
Thomas Ekman Jørgensen

Volume 3
Changing the World, Changing Oneself: Political Protest and Collective Identities in West Germany and the U.S. in the 1960s and 1970s
Edited by Belinda Davis, Wilfried Mausbach, Martin Klimke, and Carla MacDougall

Volume 4
The Transnational Condition: Protest Dynamics in an Entangled Europe
Edited by Simon Teune

Volume 5
Protest Beyond Borders: Contentious Politics in Europe since 1945
Edited by Hara Kouki and Eduardo Romanos

Volume 6
Between the Avant-Garde and the Everyday: Subversive Politics in Europe from 1957 to the Present
Edited by Timothy Brown and Lorena Anton

Volume 7
Between Prague Spring and French May: Opposition and Revolt in Europe, 1960–1980
Edited by Martin Klimke, Jacco Pekelder, and Joachim Scharloth

Volume 8
The Third World in the Global 1960s
Edited by Samantha Christiansen and Zachary A. Scarlett

Volume 9
The German Student Movement and the Literary Imagination: Transnational Memories of Protest and Dissent
Susanne Rinner

Volume 10
Children of the Dictatorship: Student Resistance, Cultural Politics, and the 'Long 1960s' in Greece
Kostis Kornetis

Volume 11
Media and Revolt: Strategies and Performances from the 1960s to the Present
Edited by Kathrin Fahlenbrach, Erling Sivertsen, and Rolf Werenskjold

Volume 12
Europeanizing Contention: The Protest Against 'Fortress Europe' in France and Germany
Pierre Monforte

Volume 13
Militant Around the Clock? Left-Wing Youth Politics, Leisure, and Sexuality in Post-Dictatorship Greece, 1974–1981
Nikolaos Papadogiannis

Volume 14
Protest in Hitler's 'National Community': Popular Unrest and the Nazi Response
Edited by Nathan Stoltzfus and Birgit Maier-Katkin

Volume 15
Comrades of Color: East Germany in the Cold War World
Edited by Quinn Slobodian

Volume 16
Social Movement Studies in Europe: The State of the Art
Edited by Olivier Fillieule and Guya Accornero

Volume 17
Protest Cultures: A Companion
Edited by Kathrin Fahlenbrach, Martin Klimke, and Joachim Scharloth

Volume 18
The Revolution before the Revolution: Late Authoritarianism and Student Protest in Portugal
By Guya Accornero

Volume 19
The Nuclear Crisis: The Arms Race, Cold War Anxiety, and the German Peace Movement of the 1980s
Edited by Christoph Becker-Schaum, Philipp Gassert, Wilfried Mausbach, Martin Klimke, and Marianne Zepp

Volume 20
A Fragmented Landscape: Abortion Governance and Protest Logics in Europe
Edited by Silvia De Zordo, Joanna Mishtal, and Lorena Anton

Volume 21
Hairy Hippies and Bloody Butchers: The Greenpeace Anti-Whaling Campaign in Norway
Juliane Riese

www.ingramcontent.com/pod-product-compliance
Ingram Content Group UK Ltd.
Pitfield, Milton Keynes, MK11 3LW, UK
UKHW021839100926
8556IPUK00004B/429